THE WHOLE TRUTH

ALSO BY T. COLIN CAMPBELL, PHD

The Future of Nutrition

The Low-Carb Fraud

Whole

The China Study

THE WHOLE TRUTH

The Hidden Agendas Behind Decades of Health Misinformation—and What to Do About It

T. COLIN CAMPBELL, PHD,
AND NELSON CAMPBELL

BenBella Books, Inc.
Dallas, TX

This book is intended to contribute to public discourse about nutrition policy and scientific integrity. It does not constitute medical, nutritional, or professional health advice. Readers should consult qualified healthcare providers before making dietary or lifestyle changes. The views expressed are the author's own and are based on his interpretation of scientific research and personal experience. The author and publisher assume no liability for actions taken based on information in this book.

BenBella Books, Inc.
8080 N. Central Expressway
Suite 1700
Dallas, TX 75206
benbellabooks.com
Send feedback to feedback@benbellabooks.com

BenBella is a federally registered trademark.

Printed in the United States of America
10 9 8 7 6 5 4 3 2 1

Library of Congress Control Number: 2025053329
ISBN 9781637747551 (hardcover)
ISBN 9781637747568 (electronic)

Editing by Leah Wilson
Proofreading by Martha Gallant and Rebecca Maines
Text design and composition by Aaron Edmiston
Cover design by Sarah Avinger
Cover image © Adobe Stock / Frogella.stock (paper curl) and Unsplash / Jacopo Maiarelli (vegetables)
Printed by Lake Book Manufacturing

CONTENTS

PART 3: THE CHANGE
(NELSON CAMPBELL)

INTRODUCTION

Now that my life has slowed down a bit, I have had more time to reflect on how I spent my last ninety-one years on this planet and on the lessons I learned. As I do this, I find myself wandering from my research into larger ideas relating to politics, economics, and ethics. I decided to write this book with my son, Nelson, to share some of these ideas, which I think you will find not only interesting but actionable.

Just a few months after I began writing in late 2023, a highly regarded health policy think tank, the Commonwealth Fund, released an ominous report on the state of American healthcare. This report was written by a panel of research scientists with access to the healthcare statistics of ten high-income countries, including data from the World Health Organization and similarly reliable sources. They found that, of the ten national populations studied, "Americans die the youngest and live the sickest lives, despite the U.S. spending the most on healthcare."[1]

I understand this sad statistic well, after spending seventy-plus years of my career in academia 1) teaching biochemistry and nutrition in classroom settings, 2) directing a large NIH-funded experimental research group mostly investigating nutritional biochemical features of human health, especially concerning cancer,

3) lecturing at medical schools in more than forty U.S. states and in more than thirty countries, and 4) being a member of several expert committees developing state, national, and international food and health policies. In these seventy years, I have also seen firsthand how vital information on our health has been overlooked and suppressed. Even the aforementioned Commonwealth Fund report failed to mention that the most important determinant of our health is nutrition.

Throughout my career, I found sheer joy in the act of pursuing scientific truths, especially relating to the ways our bodies naturally create health when provided with proper nutrition. In my research, teaching, and policymaking, I followed the evidence, regardless of how it might have conflicted with established assumptions and beliefs. While doing so, I was ostracized and demonized by many sectors of the medical establishment, who did not care for my questioning these age-old assumptions upon which their careers were created and sustained.

This book is my story of scientific discovery and of the reactions to these discoveries within institutional settings. In sharing this, I hope to show how our healthcare system often works to keep us sick, corporations profitable, and politicians and government bureaucrats in power. When the public is less informed on matters of health, the opportunity is greater for individuals and institutions with power to take advantage of those with less power, feeding people the wrong foods and promoting toxic chemicals as cures. Informing the public about health, and the ways in which they have been manipulated, is my purpose in writing this book.

In *The China Study*, I told the story of modern-day medical research and clinical practice, which gained a foothold in the late 1800s, when medical practice was being formalized around what was considered science-based research. Then, in *Whole*, I offered some thoughts on a more wholistic view of biology and, in *The Future of Nutrition*, what this might mean for the future. But still, even though I have written and lectured widely and have witnessed growing awareness of the benefits of plant-based nutrition, I remain disappointed with the pace of change and my limited success in being able to adequately explain to a wider audience the importance of nutrition to human health.

I could write an academic dissertation on why, decades after my most pivotal research, the world is far from plant-based. But rather than giving another seminar, I thought it might be more effective to tell some stories from my journey that help to explain the forces that continue to suppress the most vital health truth of our age. I think you also may find this story format more enjoyable to read.

I have hesitated to share my account until now, partly because I was never quite sure how best to tell it. I did not want to risk offending anyone unnecessarily. Nonetheless, I feel more compelled than ever to help shine a light on what is happening—even if it means stepping on some toes—so that we can wake from our slumber to build a healthier world.

PART 1

MY JOURNEY TO THE TRUTH

CHAPTER 1

A LIFE OF TWISTS, TURNS, AND LESSONS

This is a book about truth—why we often fail to see it, and worse, purposefully communicate falsehoods that hide it. It's an old saying that *truth can be a stubborn thing; it doesn't go away*. Discovering and communicating truths that can change our world, however, can be a long and painful process.

This is the ultimate idea I want to examine in this book. Unlike in my past writings, where I wrote with the voice of an educator, this time I have leaned more into the role of storyteller. After all, my life has been one long story. I have learned much along the way, not only from my education and research, but also from my difficult experiences challenging the status quo. I would like to share stories of these experiences, which all relate to nutrition but touch on deeper and more universal ideas about truth.

This book has three parts. I will tell the story of my research first, which I have done many times in other venues and formats previously. I want to offer this one more time, however, so that you understand the role of nutrition in your health. You'll then know the outrage I felt in the stories I will tell in the second part of the book.

These are stories about people and organizations who have misled the public on matters of health. On its own, each story may seem incidental and perhaps merely personal. However, when we consider them collectively, commonalities emerge. Corrupt behavior often occurs out of sight of the public but still influences the public's understanding of human health. Second, this behavior usually serves an economic, political, or reputational purpose.

I have ordered these stories to follow a larger arc, starting where we see powerful people attempting to define "truth" for the rest of us. I will then show how the media amplifies such supposed "truths," and once they become ingrained into the public consciousness, how they find their way into public policy as well as the windfall that corrupt policies can provide to powerful industries. All of this will build to my final two stories, which show the full depths to which some of us have sunk. These are stories about how we take advantage of the people in our society who are most defenseless—our children.

But I don't want you to finish this book feeling angry or hopeless. I think it's important we see the darkness, but only so that we can find our way to the light. To help with that, I will hand over part three of this book to my oldest son, Nelson, who has been thinking for most of his adult life about the ideas of wholism, reductionism, and social responsibility, not only with respect to human health but also in the spheres of philosophy, spirituality, economics, the environment, and politics. Nelson is well suited to offer some closing ideas. He has a long history working in this field as a filmmaker, nonprofit leader, businessperson, and advocate of the plant-based nutrition message, and by virtue of these experiences, he has some enlightening and inspiring ideas to share.

I think sometimes we wander through life without reflecting enough on where we have been. When we look back, we are looking with the knowledge that comes from hindsight, and it's this more informed perspective that enables us to draw life lessons that point us in the right direction today. As I write this, at ninety-one years of age, I reflect constantly on my life and feel that I can now see and understand more than I have ever seen and understood.

I would like to share some of this with you and will start by telling the story of my early life, then my later research. Many people think of me as a scientist who helped validate the health benefits of a plant-based diet, but I think my deepest understanding—an awareness that first began taking form during my childhood on a farm—is about the way we see. This may sound a bit mysterious, but you'll understand soon.

Let's start at the beginning.

THE EARLY YEARS

It was March 14, 1934, when I first saw the world, supposedly delivered by a stork on a tabletop in Annandale, New Jersey. I was the first child in my family, followed by two brothers and one sister. Now I wonder how all those years could have flown by so quickly.

Our dad was an immigrant from Northern Ireland who first set foot on American soil when he arrived on Ellis Island as a young boy with his family. Several years later, he received his first job from a wealthy man who needed a chauffeur. Like many people chasing the American dream, my father believed in hard work, so it was not surprising that he ended up impressing his boss. What was surprising was that his boss showed his gratitude by gifting my father the money to fund his dream of a small farm. That's how we landed on a hundred-acre dairy farm near Lebanon, New Jersey.

One of my earliest memories is from around 1937, when I was just three years old. I can still vividly picture a cow in our barn angrily attempting to gore me with her horns. It happened during milking time. When my parents milked the cows, my mother would put my brother and me in a big wooden bin of loose cow feed to play. That day, I'd climbed out of the bin and started walking toward the milking area. At the same time, my dad had just separated a nursing cow from her calf to bring her into the milking area. The cow didn't want to leave her calf and got loose from my father. She raced my way, hitting me in the chest and then tossing me into the air. When I landed on the concrete floor, she pulled back and cocked her head, preparing to stab me with one of her horns, which very likely would have killed me. Thankfully, my dad was there with a heavy push broom, slamming the cow on the head so hard that it broke off one of her horns and sent it flying, giving my dad a chance to grab me and take me to our house, about a hundred yards away. I remember peeing my britches, but I also remember the lesson that you don't mess with a cow when she thinks someone's threatening her greatest treasure—her calf.

My parents decided to move from that farm in New Jersey during the fall of 1941, leaving care of the cows to my father's younger brother, Tristram. We headed to Northern Virginia, where my father helped his older brother, Jack, start a construction business.

At the time, my Uncle Jack's wife, Aunt Ethel, was dying of cancer in a Washington, D.C., hospital. We lived in the same farmhouse with Uncle Jack, who often brought me and my brother (also named Jack) to visit Aunt Ethel. That was my first introduction to cancer, and I quickly learned just how serious an illness it was, as my aunt grew sicker and sicker and soon passed away. This impacted me deeply, and I frequently wished during my growing-up years to find a cure for cancer.

We returned to our farm in New Jersey a couple of years later, but after I finished third grade, my dad sold the farm and moved back to Uncle Jack's farm near Purcellville, Virginia. It was there that I attended fourth grade in a three-room schoolhouse nestled in the foothills of the Blue Ridge Mountains, where many of the students went barefoot until late fall. Though I don't remember what I learned, I have vivid memories of my time there. We spent time outside pulling weeds from the flower bed, cleaning the grounds, climbing rope, and using an outhouse—or turning it over, once with a classmate in it! We tried not to get into too much trouble because our teacher spanked kids with a wooden paddle when they misbehaved (only boys, as I recall).

After a brief stay on my uncle's farm, we moved to a new farm my dad bought near the tiny hamlet of Waterford, Virginia, where I attended grades five through seven in another three-room school. Until high school, my next youngest brother and I were expected to be in the barn every morning by 4:30 AM to milk the cows, and then to perform other chores assigned to us before and after school hours. We didn't question this routine—it was expected life on the farm—but in later years I realized how hard we had worked compared to others our age. I wouldn't have wanted to grow up any other way, though, because I learned how to work hard and with care, which served me well throughout my later life.

During the summers from the mid-1940s to mid-1950s our work progressed to a higher level. From 1944 to 1946, my brother and I tended the "bagging" machine on a small tractor-pulled combine used to thresh grain. We then started working under contract with local farmers, driving a big self-propelled combine. This was a big deal because of the large size of those machines (for those days) and because we had the first such machines in our county. It was also a big deal for us because it provided the money we needed to pay our college tuition.

Looking back, life on our farm was idyllic. We spent most days outdoors, making hay, harvesting grain, plowing fields, collecting eggs, chopping wood, and tending to animals. It was just us and Nature. In whatever spare time I had, I helped my mother in her garden, fished the stream in our meadow for sunfish, bass, and eels, hunted squirrels, rabbits, and deer, and rode my horse, Smoky. I only knew farm life and thought I would be working outdoors for the rest of my life.

Because our father only had two to three years of elementary schooling, he wanted me and my siblings to get the education he missed. This meant not sending me to the local high school in nearby Leesburg, where only about 10 percent of graduates went on to college. Not having money for a private school, he researched prestigious public schools and found one that was perfect—but located in Washington, D.C., a hundred-mile round trip from our country farm.

My father was not deterred. Until I turned fourteen and got my driver's license, he drove me to that school every day, while also working at a cinderblock factory to earn additional income. I graduated in 1952 from Western High School, named by a national educational assessment organization during one of those years as the number one public high school in the U.S. It is now named the Duke Ellington School of the Arts.

A FARM BOY AT CORNELL

Our country is one built by immigrants seeking opportunity and by enslaved people seized in distant lands and brought here against their will. The story of my father's family is a classic Irish immigrant story. They dreamt a dream big enough to climb on a boat to sail in risky conditions across a vast ocean. My father's piece of this dream was to provide the education to his kids that he lacked.

His dream became manifest the day I received my acceptance letter to Penn State. The first in my family to attend college, I decided to major in general agriculture, then changed to pre-veterinary medicine after one year. I enjoyed college, loving the interaction with my professors and with my classmates. I quickly made lots of new friends on campus.

Along the way, I impressed my faculty advisor, Professor Bob Swope, who unbeknownst to me arranged for me to interview for early admission to two veterinary schools after only three years at Penn State. I interviewed for admission to the University of Pennsylvania and a newly established vet school at the University of Georgia. I was admitted to both, but chose Georgia.

I thought I was headed back to the farm as a large animal vet, which felt like a natural outcome of my studies. But then, toward the end of my first year of vet school in 1953, I received a surprise telegram from a well-known professor at Cornell, Clive McCay, offering me a full scholarship to study nutritional biochemistry in Cornell's graduate school. Cornell? Really? I never knew why he sent the telegram, although I suspect that this too was thanks to Professor Swope.

I completed my MS degree in nutritional biochemistry at Cornell within a year, then my path took another unexpected turn. I received a request to meet my local draft board for mandatory military service, as I had exhausted my military deferments during my years in school. Based on my Reserve Officers' Training Corps (ROTC) participation at Penn State, I was granted second lieutenant officer status.

But before reporting for my assignment on August 1, 1958, in Fort Collins, Colorado, I received yet another surprise request, this time from Cornell professor Richard Warner. He offered me a scholarship to return to Cornell for a PhD program, if possible before my military service. The military authorities once more agreed to defer my assignment until I finished the degree.

I finished my PhD in nutrition at Cornell in 1962, along with two minor fields of study (biochemistry and microbiology). My PhD research dissertation tested the ability of biuret, a nitrogen-rich chemical, to increase the growth and milk production of cows and sheep that consumed it. This research objective made sense to me; I had grown up on a farm and had always thought protein was the king of all nutrients and livestock farming was at the apex of agriculture. My results were inconclusive, although I was able to observe how the gut biome rapidly responds to nutritional intake.

MY EARLY CAREER

As is the case with many young people, I had feet inclined for wandering. At Cornell, I made friends with an Irishman named Pat Fox, and together, we hatched a dream to travel the world by working our way from one ocean vessel to another. Around the time of my graduation from Cornell, I shared this dream with my father, who promptly put an end to it. He rightly pointed out that I had outstanding loans from my education that needed to be paid.

Following his advice, I found a job at a small company in Herndon, Virginia—Woodard Research Laboratory, near my hometown in Northern Virginia and very near the soon-to-be-completed Dulles Airport. My decision to work at Woodard had lifelong implications. Upon my arrival, I was paired with a secretary, Karen Lee, who quickly caught my eye, and then my heart. We married on September 1, 1962, and now have five children, eleven grandchildren, and two great-grandchildren.

As the Vietnam War intensified, I expected to hear from the military to begin my service, but for reasons I am still unsure of,

I never heard from them. Instead, I began my work at Woodard, leading a project to test the toxicity of a synthetic hormone-like chemical intended to increase the growth of strawberries. As I considered the structure of this chemical, I surmised that, in the presence of sunlight, it could chemically break into an estrogenic product listed by the Food and Drug Administration (FDA) as a chemical carcinogen. Sure enough, when I exposed it to sunlight, it degraded into that byproduct. I can only wonder about the havoc that chemical might have caused in future years had it been approved for the marketplace.

A second project involved our lab joining with eleven other labs and the FDA to identify a mysterious chemical in poultry feed that had reportedly killed a million poultry in southern Ohio. This toxin was called "chick edema factor" because it caused a lethal accumulation of fluid in the heart sac of chickens and turkeys that consumed the contaminated feed. Within a few months, the FDA official in charge of that project, Leo Friedman, accepted a senior professorship to direct a new toxicology program at MIT, and he invited me to join him to set up his new research laboratory. I accepted and started my MIT research associate position in Cambridge, Massachusetts, in August of 1963.

Friedman wanted me to chemically isolate this unknown and highly toxic chick edema factor. Within two years at MIT, I succeeded in producing a nearly pure form of the chemical, and I published the isolation procedure in 1966.[2] This paper caught the attention of the U.S. government, because this same chemical was in Agent Orange, sprayed by the military in Vietnam to defoliate forest cover in areas used by North Vietnamese troops. A year or so later, government researchers identified its structure and named it "dioxin."

The part of poultry feed that contained the dioxin—an oil—was said to come from crop plants sprayed with the synthetic plant hormone 2,4,5-T to make them grow faster and bigger. The knowledge that dioxin was in a spray used on crops ultimately for human consumption was an early education for me in the chemical contamination of our food supply.

According to calculations from my isolation procedure, it appeared to be one of the most toxic chemicals ever isolated, even more deadly than botulism toxin. Unfortunately, I experienced considerable exposure to this newly isolated chemical because I had worked in an old lab without a protective chemical hood, leaving me to breathe in heavy amounts of it, as well as the organic solvents used to isolate it, on a nearly daily basis. I wasn't the only one to experience this; a colleague at an FDA laboratory working on the same toxin became ill and died at age forty-one.

For twenty-five to thirty years after this work, I experienced symptoms that included intermittent migraines and polyps in my sinuses that had to be cauterized. In the late '70s, I started feeling numbness in my facial and neck muscles, which became increasingly severe through the 1980s. I eventually lost most control of these muscles, ending up barely able to eat or to speak. Around this time, dioxin levels in my bloodstream were measured at over eight hundred times the allowable safe limit.

I stepped out of public view when I could no longer lecture and prepared for the worst. But then I had the good fortune of a chance encounter with a group of naturopathic doctors, who suggested I try a medically supervised fast. This made sense to me because dioxin accumulates in fatty tissues, which I would lose during the fast. I underwent two such fasts within twelve months and, lo and behold, was able to expel most of the dioxin from my system. Karen and I also got very strict with our diet, and in the subsequent years I slowly regained control of these muscles as the damaged areas of my nervous system rewired and healed.

After my time working at MIT, I was recruited to a faculty position at Virginia Tech in their department of biochemistry and nutrition. This next chapter proved to be a major turning point in my life, when I had to confront a powerful bias I had learned from my early years on a farm and later reinforced in my higher education.

CHAPTER 2

AN UNEXPECTED DISCOVERY

We were only moving down the East Coast, from the Boston area to southwest Virginia, but it felt to Karen and me like another major life change. Fortunately, we did not have much to move and had only one child, Nelson, who had found his way into the world just a few months prior. Together, we moved to a small white house not far from the university, nestled in a middle-class neighborhood in Blacksburg.

I started my work at Virginia Tech in a research lab, using a fungal culture to synthesize a supply of an extremely toxic natural chemical called aflatoxin.[3] This chemical is produced by a fungus, *Aspergillus flavus*, which grows on food products like peanuts, cottonseed meal, and corn. Initially documented by researchers in England, its chemical structure was determined at MIT in the same laboratory where I had worked on the dioxin project. Aflatoxin causes primary liver cancer and has long been considered a potent chemical carcinogen.[4]

My primary early focus at Virginia Tech, however, was not my work on aflatoxin but was instead an opportunity I had never envisioned. Professor Charlie Engel, who chaired the department and had recruited me to my position, invited me to coordinate the on-campus side of a project he was directing to help malnourished children in the Philippines, which also required frequent visits to the Philippines to assess local program activities. The U.S. Agency for International Development (USAID) was funding the project, an agency Dr. Engel had come to know while serving in the U.S. Army in the Philippines during World War II.

The Philippines project was an extension of a similar program in Haiti involving so-called "mothercraft" centers and co-directed by Professor Kendall King in our department.[5] These centers provided assessment, feeding, and nutrition education services for malnourished children and their mothers. So, before traveling to the Philippines, I traveled to Port-au-Prince, Haiti, to see how the program had been conceived. Visiting the poorer neighborhoods of the city, I was struck by the indescribable poverty suffered by these families and their children. At that time, about one-half of the children in Haiti were dying before the age of five. Their images are burned into my memory, as vivid now as they were then, and I still feel an upwelling of grief whenever I recall them. I would come to see pockets of the same kind of poverty and childhood malnutrition in the Philippines, especially in and around Manila, where poor neighborhoods were often on the opposite side of community walls or railroad tracks from the homes of the rich.

When I took the position of coordinating the Philippines program, I assumed our primary focus should be on ensuring malnourished kids' access to high-calorie foods. According to the prevailing view of the time, this was done best using the animal-based foods commonly consumed in rich countries, which contained so-called "high-quality" protein. This made sense to me as a farm boy turned nutritional scientist.

Life on the farm, though, also taught me to see through my biases, opening me to ideas that did not fit into the prevailing paradigm. On the farm, I got glimpses of the interconnectedness, complexity, and beauty of Nature. I realized we are but a very small part of this larger whole. We are not above, or in command of Nature, and if we are in fact "commanding" Nature, those commands ring hollow—because at the end of the day we sustain ourselves only at the mercy of Nature, which must always remain in balance. I could not articulate these ideas at that time, but I felt these things to be true, and I think this knowledge of our small place in the universe fostered within me a humbler way of seeing. Rather than commanding certain things to be true, by seeing whatever "facts" reinforced my biases, I opened myself to seeing the world in front of me in whatever form it took.

One such opening occurred on a golf course, of all places, where I met two Filipino physicians who inquired about my work. Our discussion between golf swings wandered to the topic of my lab research at Virginia Tech on liver cancer and its initiation by aflatoxin, at which point they shared that there was evidence suggesting that kids from more affluent families in the Philippines, who were consuming a more animal-centric diet, had higher rates of cancer and other Western diseases compared to kids from poorer families. One of the physicians worked at a high level, advising the national government, so while I could not directly document this claim—and knew that the kids from poor families also received less medical care and, therefore, fewer diagnoses—this story from a seemingly credible source nonetheless got me thinking.

This thought process deepened when I discovered around the same time a paper reporting the findings of an experimental study in India. This paper reported that animal protein dramatically increased aflatoxin-initiated liver cancer in lab rats.[6] I started to wonder if animal-based foods really were the healthiest option for the children in our program, and if there wasn't some piece of information I was missing.

A light bulb was starting to flicker on in my head and I wanted to investigate this diet–cancer connection further, most of all because we were being encouraged to increase animal protein consumption among malnourished children. I could not accept that we might be giving them the wrong foods.

BREAKING THROUGH THE PROTEIN PARADIGM

Protein was one of the first nutrients to be isolated and named, in 1838 by G. J. Mulder and Jöns Jacob Berzelius, respectively. Its name was taken from the Greek word "proteios," meaning "of prime importance," because it was believed that protein was crucial for the body and for health.[7]

Since its discovery, protein has always been associated with eating animals. And in my earlier years, I also thought animal-based foods were important to consume for their protein content. The milk we produced on our dairy farm was assumed by everyone to be important, especially because we thought milk provided calcium and protein for bones and teeth. A few years later, my doctoral research at Cornell was premised on a perceived need to increase the production of animal protein.[8] And my trainee teaching experience was a one-credit undergraduate course called "Livestock Feeds and Feeding," which promoted protein consumption to optimize growth and production of meat, milk, and eggs.

This trajectory made sense. I was living in a paradigm telling me that animal protein was the king of all nutrients, and I was a farm boy after all. As I said, though, my years on the farm, immersed in Nature, also instilled in me a kind of humble curiosity, so when I saw facts that seemed to contradict what I was thinking at that time, I felt the urge to know more.

I went back to Virginia Tech motivated to understand this diet–cancer connection using conventional mechanistic research strategies. I had been trained to think about biology as a mechanistic system, where every outcome could be explained through linear causation, and if we could identify the key mechanism of action, then perhaps we could change the outcome. Indeed, most research is still done this way today, a topic I will return to a little later.

With this mindset, I set out to find a mechanism of action that might explain a causal connection between diet and cancer. Along with a team of research assistants, I looked at factors like the entry of carcinogens into cells, their processing by mixed-function oxidase enzymes, the metabolites produced by those enzymes and their genetic impacts, DNA repair function, cell replication, reactive oxygen radicals, cancer-promoting hormones, natural killer cells, and on and on.

One consistent theme emerged in our work. No matter what mechanism we examined, it seemed that animal protein was always impacting that factor in a way that initiated or pushed the cancer process forward. The more we looked, the more certain I became of the causal connection between animal protein and cancer, but also the more uncertain I became of any one mechanism of action being the most important.

I should note that in all these years of research, I never took money from corporate interests. Instead, taxpayers funded our research. Upon returning to Virginia Tech from the Philippines, I applied for a grant from the National Institutes of Health (NIH), who rigorously reviewed the proposal and awarded the funding. I submitted eleven more applications over the ensuing thirty years, all approved. Whereas the average success rate for other funding applications was only 15–16 percent, we obtained funding for all our applications, solely based on professional scientists judging them to be of high quality.

I mention this to underscore that our research during these years was done using good design and in the most rigorous manner. Yet, since that time, our research findings have been rarely professionally cited, even though they were published in top-of-the-line biomedical journals. For those people promoting the production and consumption of animal-derived foods, this research is best left alone, abandoned to rot in the digital universe. Fortunately, much of it is still accessible and is listed in my bibliography, published on our Center for Nutrition Studies website.[9]

I spent a total of ten years at Virginia Tech, 1965–1975, where I organized and led research projects resulting in the publication of well over a hundred peer-reviewed research papers. This work began attracting considerable attention, including from researchers at Cornell University.

BACK TO MY ALMA MATER

It's interesting how life can sometimes take us in a circle. We often end up going through various life experiences only to learn we were never traveling in a line, but in fact along a curved path taking us to a place that feels strangely familiar.

This happened to me in early 1975 while I was still at Virginia Tech. I was approached by the director of the Division of Nutritional Sciences at Cornell University and offered a full professorship. While impressive, this offer came without the usual tenure and "academic freedom" clause that often goes with this position. I had earlier earned this privilege at Virginia Tech in 1969, when I was promoted from assistant to associate professor. Tenure means having a full professorship for life, wherein one cannot be fired for expressing opinions and conclusions, whether in public discourse or in the classroom. I did not accept the offer initially, at which time they sent it again, this time with the tenure clause included.

This was perhaps one of the best career decisions I ever made. I came to Cornell bringing with me the NIH funding already awarded, a couple graduate students, and my continued engagement in research and professional activities off campus—expert panels, policy development activities, and research presentations.

Because my research had been challenging the official dogma of the day, I also brought with me growing controversy. Had I not been granted tenure, it would have been difficult for me to pursue this science in the honest way I sought.

One example of this was the laboratory research I conducted suggesting that cancer could be turned on or off merely by toggling between plant and animal protein, a finding supporting the results of the earlier study in India that I learned of while in the Philippines.[10] This effect was unusually convincing—an all-or-nothing type of response. It occurred rapidly and worked through multiple synchronous biological mechanisms. This research finding called into question the popular concept that cancer is primarily a genetic disease and qualified animal protein as a potent carcinogen.

It was also at this time, early in my tenure at Cornell, that I began to catch glimpses of the most radical of all the truths I have discovered in my life. As I noted earlier, in the years leading up to this point I had been conducting research through a reductionist lens, always looking at a particular mechanism of action and how that particular factor might be linked to nutrition and cancer. The more of this research I did, the more I realized there is no single mechanism of action within the cancer process that is more important than the others. I started to understand more deeply the complexity of biology and that my research career had largely been based on a false premise: the idea of mechanistic biology. I'll be returning to this important topic later.

Up to this point, our research on animal protein was mainly based on experiments with rats and mice. But was it relevant to humans, too? A unique opportunity to explore this question arose following President Nixon's landmark visit to China in 1972 and the subsequent thaw in the relationship between China and the U.S. In 1980, a senior Chinese delegation visited the U.S., which included Dr. Junshi Chen, who was serving as deputy director of the Institute of Nutrition and Food Hygiene in the Chinese government's Academy of Preventive Medicine.

Dr. Chen's family was well-known in China. His maternal grandfather was the intellectual partner of the revolutionary Sun Yat-sen, who overthrew the centuries-old Qing dynasty in 1911. Dr. Chen's father was the first Chinese ambassador to the United Nations in 1946–1949.

Trained in medicine, Dr. Chen also had professional experience in nutrition. He was leading a research team investigating an association of selenium deficiency with Keshan disease, a childhood heart disease. He also had interest in diet and cancer, so he contacted my office while in the U.S. to inquire about visiting our campus. I jumped at the opportunity and granted him almost a year in my lab to continue his research studies on the effect of selenium on the body's ability to cleanse itself of carcinogenic toxins.[11]

From Dr. Chen, I learned of an immense atlas of diet, lifestyle, and disease mortality in nearly 2,400 Chinese counties just then being published in China.[12] Mortality rates for about fifteen types of cancers and thirty-five other diseases were widely divergent in different counties, suggesting that local dietary and lifestyle conditions, rather than genes, were likely causative. I learned, too, that Dr. Chen personally knew the director of this atlas, Dr. Junyao Li, who was in the U.S. at that time visiting the National Institutes of Health (NIH).

I asked Dr. Chen whether he and his team might consider doing a joint research project to investigate the possible causes of these well-documented cancer deaths, and he agreed. I then sought collaboration with the world's best-known epidemiologist, Sir Richard Peto at the University of Oxford, who also agreed. This became the first modern-day joint research project between the U.S. and China. On a more personal level, it also allowed me to create treasured relationships with Drs. Chen and Li, Sir Richard, and many of their colleagues, who visited my lab for weeks and sometimes months at a time.

Funding for this project, conducted in 1983–1984, was provided by the U.S. National Cancer Institute (of NIH).[13] Additionally, China provided survey teams who collected and recorded data from 130 villages, comprising 6,500 adults between thirty-five and sixty-four years of age and their families. A second, even larger survey of 8,990 adults—including thirty-two areas in Taiwan—was completed in 1989, the first modern-day research project that included both mainland China and Taiwan. I cannot praise highly enough the competence and the diligence of the individuals with whom I worked, both senior leaders and the healthcare workers under Dr. Chen's supervision in the Chinese provinces who collected the data.

This data collection included 367 items of information, yielding around one hundred thousand statistical correlations, of which about eight thousand were statistically significant. Any one of these statistically significant correlations by itself could only be claimed to indicate an association, not a causative link. But when we saw many of the correlations pointing in the same direction—that animal-based foods, displacing plant-based foods, were associated with cancer and other chronic conditions—it became clear that we were looking at more than "association." We also observed that these diseases tended to cluster in the same geographic areas, where people shared a similar culture and ate similar foods.

The design of this study allowed us to see these larger patterns. As I shared earlier, I had begun to understand through my lab research that biology is complex, far too complex for there to be any single biological mechanism of action that can explain a disease outcome. In our research on China, I wanted to cast a wide net, looking at a large number of factors in the hope that we might see larger patterns illuminating the relationships between diet and disease.

Upon the release of our first data, a 1990 *New York Times* feature article called our study the "Grand Prix of epidemiology."[14] And in 1997, the Chinese government awarded the project first prize as the most important medical research project in China in the twenty-year post–Deng Xiaoping era, 1978–1997.

The China study was the crown jewel of my career, and as nice an ending as this would be, my research story does not end here. As I became more convinced of a causative relationship between diet and disease, I began scouring the research literature further. I learned of a large body of research performed in prior years that pointed in exactly the same direction I was now facing.

RESEARCH IGNORED AND FORGOTTEN

The total wealth of human knowledge is beyond our comprehension, even with artificial intelligence now on the scene, because so much of what people have explored and discovered has been lost to history. Oftentimes it's the power brokers in society who decide what's remembered and what's forgotten. I have found this to be true in my own profession and have often been motivated to search for knowledge long forgotten.

When I began to see more clearly the impact of nutrition on our health, I started searching for other research that might be suggestive of this same truth. I ended up going deep, not only into the prior century, but even back to the ancient Greeks. And everywhere I looked, I found discussion and evidence that confirmed what I was learning. I could write a whole series of books on what I discovered, but for our purposes here, I will simply share some more recent research on diet and disease associations across countries. The graphs shown on the following pages come from studies originally published between 1959 and 1999, and considered collectively, they are truly exceptional.[15] (I superimposed regression lines such that an equal number of data points fell on either side of each line.)

For those who may not be familiar with this kind of graphic display, there are two reference lines for all graphs. Each data point in a two-coordinate graph is described by two numbers, one referring to the horizontal X axis, the other to the vertical Y axis. Consider the colon cancer chart, where the data point for the U.S. is clearly shown. That point is at (275, 37), meaning that each individual consumes *on average* 275 grams of meat per day (X axis), and there are *on average* 37 colon cancer cases per one hundred thousand individuals (Y axis).

The deviation of other data points somewhat above or below the diagonal regression line is common, of course, and may represent normally varying measurements. The regression line itself represents an *average* of all the data.

Colon Cancer Incidence—Meat Intake (Animal Protein)[16]

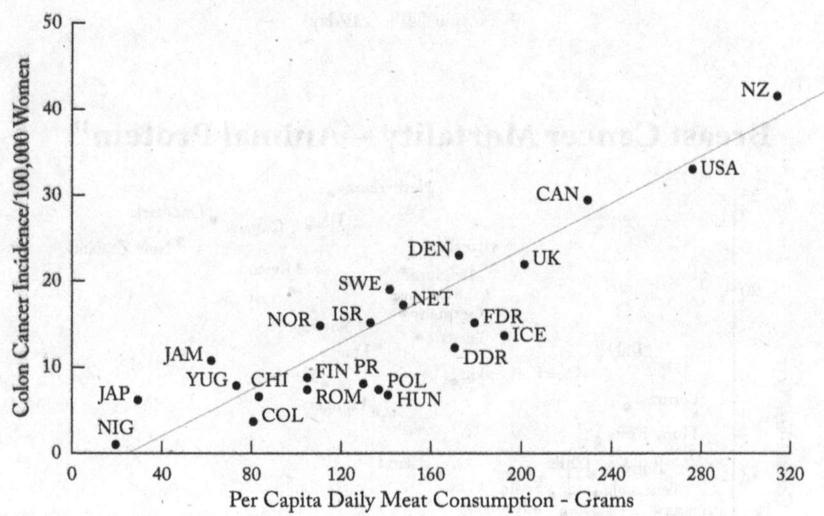

Prostate Cancer Mortality—Nonfat Milk (Animal Protein)[17]

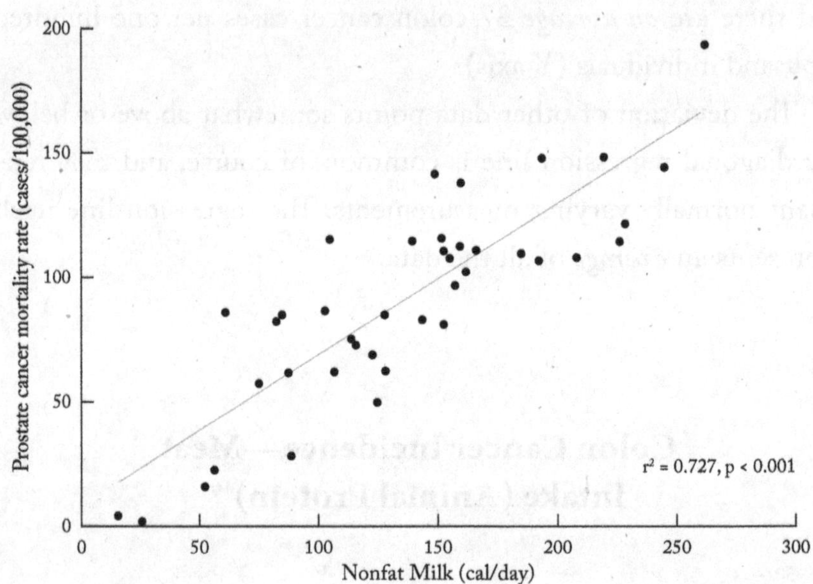

Breast Cancer Mortality—Animal Protein[18]

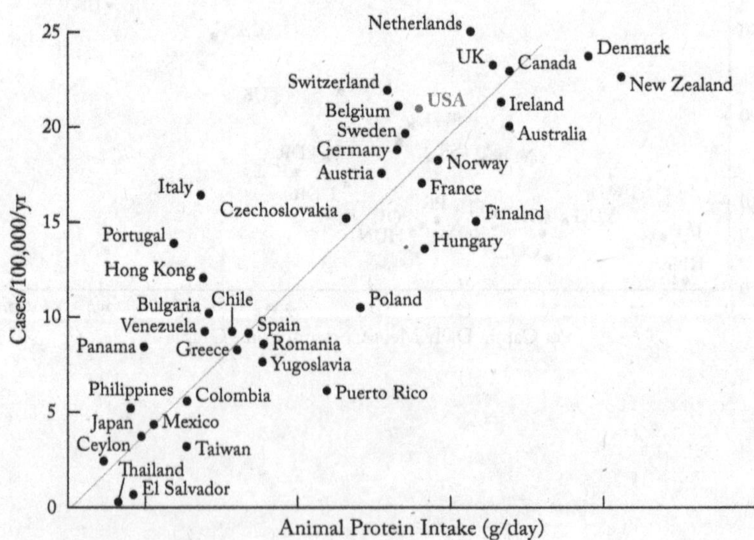

Kidney Cancer Incidence—Animal Protein[19]

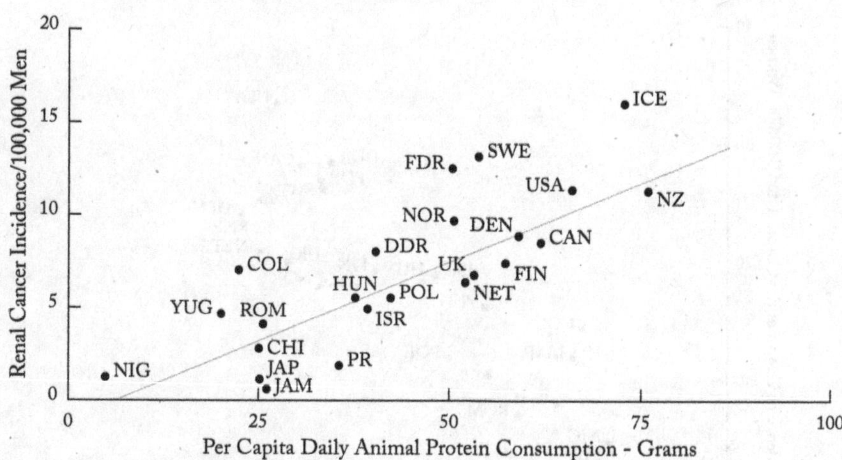

Coronary Heart Disease Mortality—Cholesterol Intake (Animal Protein)[20]

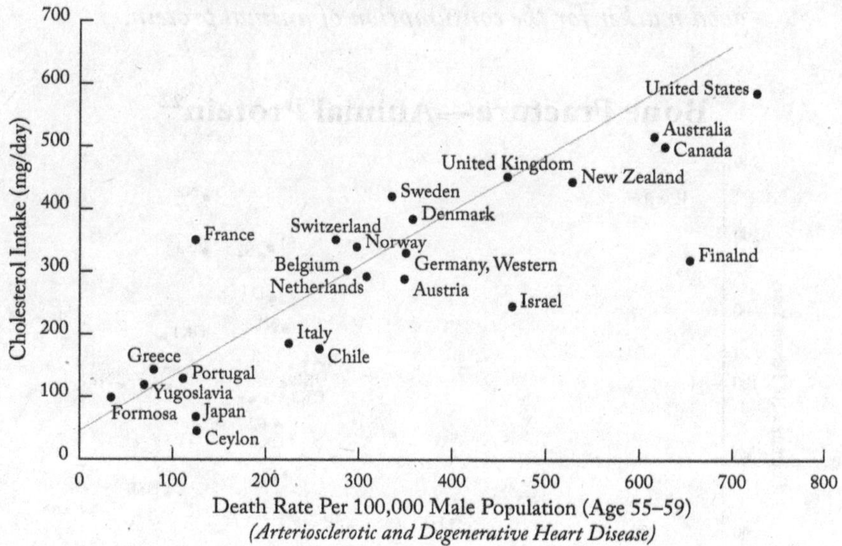

Because cholesterol is found exclusively in animal-based foods, it is a good marker for the consumption of animal protein.

Uterine Cancer—Total Fat Intake (Animal Protein)[21]

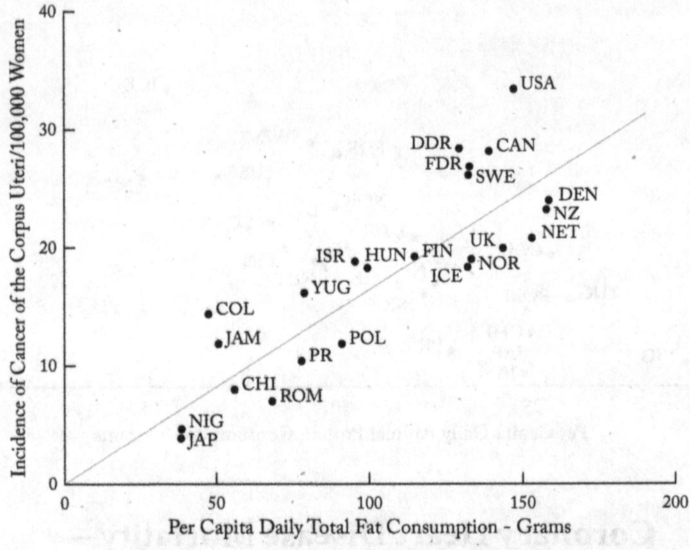

In this study, fat came mostly from animal foods, so it is a good marker for the consumption of animal protein.

Bone Fracture—Animal Protein[22]

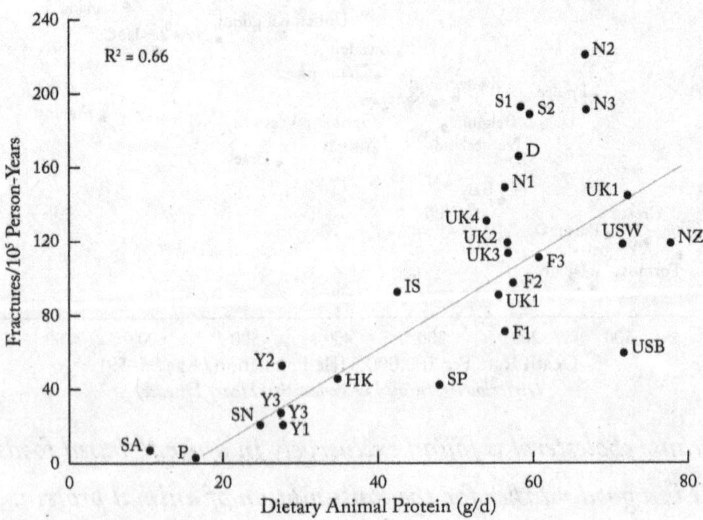

Bone Fracture—Calcium Intake (Animal Protein)[23]

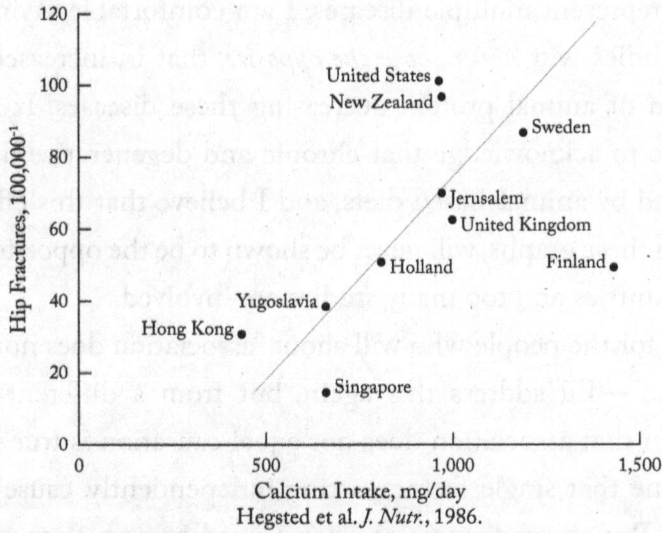

Calcium is consumed mostly in the form of dairy products, which are usually accompanied by other animal-based foods.

The regression lines for all these studies begin to increase very near the X/Y origin. This suggests that zero animal protein consumption approaches zero disease risk, supporting the hypothesis that consumption of animal protein begins increasing the prevalence of these diseases even in populations of people consuming small amounts of animal-based foods. It also is impressive that a diet richer in animal protein increases disease risk for multiple diseases that are common to economically developed Western countries. And we should not forget that increased consumption of animal protein foods also means reduced consumption of plant-based foods, and that both changes are likely contributing to the same response.

These graphs also show something else. Because these associations represent multiple diseases, I am comfortable saying that future studies will *never show the opposite*, that is, increased consumption of animal protein decreasing these diseases. It is well past time to acknowledge that chronic and degenerative diseases are caused by animal-based diets, and I believe that this effect, as shown in these graphs, will never be shown to be the opposite—too many countries and too many studies are involved.

And for the people who will shout "association does not equal causation!"—I'll address this again, but from a different angle. The claim that association does not equal causation is true only if we assume that single factors, acting independently, cause single diseases. But these diseases are not caused by one factor, unless the "factor" is *the ratio of animal-based to plant-based food consumption*, and this "factor" is not like a single enzyme or other biological mechanism—it's the seemingly infinite number of nutrients within each food group working in unison to create the same response. Note also that these cause-effect associations impressively approach linearity.

Also note the ranges of disease risk. From highest to lowest risk, the differences are tenfold, at least, and include many degenerative diseases common to Western countries all responding in the same direction. This suggests that these diseases could be mainly prevented by "one factor," a ratio of animal-based to plant-based food consumption, even though other causal factors may occasionally play roles as well.

I had arrived at a point in my journey where I knew with certainty that the consumption of animal-based foods, especially as they displace plant-based foods, is the most important determinant of our health. I went on to share this understanding in 2005 in my book *The China Study*, coauthored with my youngest son, Tom, which helped to launch the plant-based nutrition movement.

In one sense, I found myself a long way from the dairy farm I grew up on, but in another very important sense, I was circling back home through the radical way I was now seeing biology. This way of seeing, which I call "wholism," reflects what I saw in the forests, fields, and streams of my youth, scenes that are still vivid in my mind all these years later.

CHAPTER 3

WHOLISM VS. REDUCTIONISM

The most apparent attribute of Nature is its beauty. When you are in a forest, alongside a stream, feeling a cooling breeze on your face, seeing the play of sunlight among the trees, and listening to the birdsong, rustling leaves, and water cascading over rocks, you can't help but feel awe at the beauty of it all. It's in these moments, when we come face to face with a complex, interconnected natural system working in perfect harmony, that we can catch a glimpse of what I like to call wholism.

I have felt this sense of awe many times throughout decades of work, as I've peered into the complex universe that is human biology. And while my knowledge of biology has continued to grow over the years, my greatest lesson has been learning how much we do not know.

If we could see all the details within a single cell, we would feel as though we were looking into a universe. And even one tiny piece of that universe—an enzyme—is complex beyond our ability to fully understand. An enzyme is a polypeptide (a long chain of amino acid residues) folded into a complex shape that is perfectly tuned to convert chemical A into chemical B. When chemical A drifts nearby, the enzyme snaps into the precise geometry needed to carry out the conversion. While we are learning more every day about this choreography, much of it remains a mystery.

There are thousands of distinct enzymes in the body, catalyzing billions of reactions across the body every second of every day, yet we don't fully understand how a single enzyme works. Most of these reactions take place within our cells, which are universes unto themselves. A single cell contains trillions of molecules—proteins, ions, nucleic acids, lipids, and metabolites—all arranged into intricate networks of membranes, metabolism, signaling, and genetic regulation, which are in constant communication, working in unison. This complexity inside a single cell is exponentially compounded by the fact that there are tens of trillions of cells in the body, organized into tissues and organs. On a more macro scale, there are larger interwoven systems—circulatory, nervous, endocrine, immune, and others—all connected and communicating in real time, coordinating the body's trillions of cells across its tissues and organs to sustain the health of the whole.

My understanding of this aspect of biology deepened over the course of my career. In my early years, I worked within the more reductionist paradigm that still defines biological research today. I wanted to understand the single biological mechanism that might explain the connection of animal protein to cancer and other diseases. I soon came to see that it was not possible to identify such a mechanism; there were many mechanisms involved in cancer, all of them important. In a series of experiments I led with my graduate students, higher animal protein consumption (specifically casein, the main protein in cow's milk) did the following:

1. Increased transport of an initiating chemical carcinogen into the cell.
2. Increased synthesis of mixed-function oxidase, causing it to convert carcinogens into highly reactive metabolites—which in turn bind to DNA and cause mutations that initiate cancer.
3. Increased in a dose-dependent manner the formation of the carcinogen-DNA complex (adduct) that leads to mutations.
4. Decreased the ability of cells to routinely repair this adduct.
5. Decreased production of so-called natural killer cells, made by the immune system to attack infectious agents and cancerous cells.
6. Increased production of a growth hormone (specifically, insulin-like growth factor) that increases cancer cell growth.
7. Modified the metabolism of calories in a way that increases cancer growth.
8. Increased formation of chemically reactive oxygen molecules that promote cancer (and aging, among other effects).
9. Increased cell replication.

All of these biological mechanisms we observed led to the same effect—increased cancer growth. That is, increasing consumption of animal protein increases cancer growth through all of these mechanisms, apparently working together, in a way that we could not have recognized had we been focused on one mechanism alone.

Another way to see the complexity of biology is through the way food, and specifically nutrients, is utilized within the body. The potential number of nutrients is incalculable. Any one nutrient is often a member of a group of closely related substances, called isomers and analogs, all creating the same type of response but to different degrees. For example, beta-carotene, an antioxidant in plants, is only one of hundreds of related "carotenes" (carotenoids) in plants, most if not all having similar antioxidant activities, but differing in their degree of activity.

This complexity deepens through endless nutrient-nutrient interactions during digestion, absorption, transport, metabolism, storage, and excretion. We will be living in skyscrapers on Mars long before we truly understand all Nature's nutrients and their relatives, and their endless interactions and effects.

By now you should be getting a good sense of what I mean by "wholism." And with this understanding, perhaps you can see the absurdity of so much of the research and public conversation today around biology and health, which is based on a much simpler, more mechanical view of biology.

"TINKERTOY" BIOLOGY

In a world of smartphones, virtual reality video games, artificial intelligence, and other rapidly evolving technologies, many people may have forgotten the simple toys from our past. One such toy was the Tinkertoy construction set—a set of sticks and wooden round connectors that kids could use to build simple structures. I think this is a good metaphor for our modern view of biology.

Despite their advanced degrees and papers full of complex, technical language, many researchers are still looking at biology as a mechanical, grossly simplified process—in short, they are interacting with biology as kids play with Tinkertoys. They peer into an infinite web of complexity to study a single factor and how that factor might be connected to a disease. But when we look at single mechanistic factors to the exclusion of the complex whole, we can reach incorrect conclusions. This is not to say reductionist research has no value. After all, my early studies on discrete mechanisms involved in the cancer process deepened my understanding of the connection of nutrition to cancer and of the true nature of biology. However, reductionist research, divorced from a more humble and wholistic understanding of biology, has been the source of much confusion in our community.

OUR FUTURE FORETOLD IN OUR GENES

Did you know that the length of the DNA in a single average human body might be as much as two hundred and forty trillion feet? That's roughly the distance from the earth to the sun five hundred times over! And all that DNA contains information essential to who we are, how we live, and how we die. Our genes are important indeed, but alas, they are not the end-all and be-all of human biology, as we have long been led to believe.

Generally speaking, it has long been thought that cardiovascular diseases, cancers of various organs, and other chronic diseases "run in the family," meaning that they are caused by genes passed from one generation to the next. And because we can't change our genes, the thinking goes, we therefore need to find other ways to prevent and manage these diseases when they inevitably occur, such as through drugs. This belief goes by the name "genetic determinism," which sounds impressive but in reality is another Tinkertoy concept.

While genes may predispose us to a disease, whether that disease occurs or not is determined by *genetic expression*. The research I performed over my career, and the research of others (such as the international studies I shared earlier), all underscore the importance of genetic expression. Diseases develop because of the ways that environmental factors, most notably nutrition, control how, when, and what gene products are produced, usually modifying enzymes that participate in disease formation (i.e., genetic expression).

Over the years, more scientists and researchers have come around to theories of genetic expression. However, many others are still leery of the idea that nutrition can alter genetic expression in ways that significantly impact the development and growth of cancer and other chronic diseases, as my research findings have shown. The pharmaceutical industry in particular knows well that if the public were to understand the power of nutrition to control disease, it would greatly decrease the need for drugs and have serious consequences for their bottom line.

CHEMICALS AND CANCER

As with genes, we sometimes see chemicals as the chief cause of cancer—a perspective that has become even more pronounced as of late and is yet another example of a simplistic view of biology.

Before I go further, let me introduce an important qualification. While I believe there is a more important cause of cancer than chemicals—an animal-based diet—I am also opposed to the use of toxic chemicals in agriculture and processed foods, both for health and environmental reasons. There are a host of adverse health impacts from our overuse of herbicides, pesticides, and genetically modified organisms (GMOs), and from the chemical ingredients in processed foods. As I shared earlier, I know the dangers of chemical exposures personally; I spent many years recovering from the effects of dioxin on my nervous system. The impacts of synthetic chemicals on our environment are similarly concerning. We are destroying our natural world, including the microbial life essential to the health of our soils, as we drench everything with toxic substances totally foreign to Nature.

That being said, though, I think our fixation on chemicals as the most important determinant of health is actually an offshoot of the genetic determinism argument. Certain toxic chemicals modify (mutate) genes of normal cells to give them the "genetic" capability to initiate cancer, in which case they are called chemical carcinogens. Once such a genetic mutation(s) occurs, the thinking goes, the process automatically proceeds to full-blown cancer.

As I partially recounted earlier, I worked in two cancer testing laboratories in the late 1950s and early 1960s whose main business was to test and then regulate chemicals for their cancer-causing activities in experimental animals. However, the evidence that has accumulated over the decades has shown something surprising: as unnatural and obnoxious as these chemicals may be, they are not the primary cause of cancer.

This, then, poses the question: what *is* primarily responsible for cancer, the second leading cause of death in the U.S.? I think the verdict is now in—the food we eat is, specifically our consumption of animal-based foods at the expense of plant-based foods. And though I am speaking of cancer here, the same can be said about chronic degenerative diseases more generally.

HEALING IN A PILL

There is a strong motive for perpetuating these myths about genes and chemicals—and that motive is financial. If cancer is a function of only genes and chemicals, then we are helpless victims who have only one choice, which is to rely on the pharmaceutical and other technological "solutions" of our healthcare system.

The pharmaceutical industry, now earning well over $1.5 trillion dollars a year, represents the apex of reductionist thinking.

They see the world through a reductionist lens, not only because of their mechanistic understanding of biology, but also because this paradigm enables them to make more money.

Drugs are chemicals designed to act on specific metabolic targets that are thought responsible for the causation of single diseases. These chemicals are compromised on two accounts. They (1) are often synthetic, and thus are foreign to Nature, and (2) act independently in a very complex biological environment, inviting unpredictable effects. Incredibly, the drug solution so esteemed within our healthcare system is widely recognized as one of the leading causes of death in the U.S.

Drugs also don't work to resolve the root causes of disease. Consider this example. The first statin, lovastatin, was heralded by the press as a miracle cure for heart disease. So promising was the message on statins—which lower cholesterol—that some news accounts suggested they might even be added to public water supplies![24] Yet, statins have not delivered the dramatic real-world benefits that were advertised. A 2022 systematic review and meta-analysis with over 140,000 participants showed that, over an average 4.4-year follow-up, statins produced absolute risk reductions of only 0.8 percent for all-cause mortality, 1.3 percent for heart attacks, and 0.4 percent for stroke.[25] So little benefit, yet so much money earned. The statin market now exceeds $16 billion.[26]

Compare this to the efficacy of a whole food, plant-based diet, as shown in the graph on the next page.[27] I've added a "best fit" linear regression line.

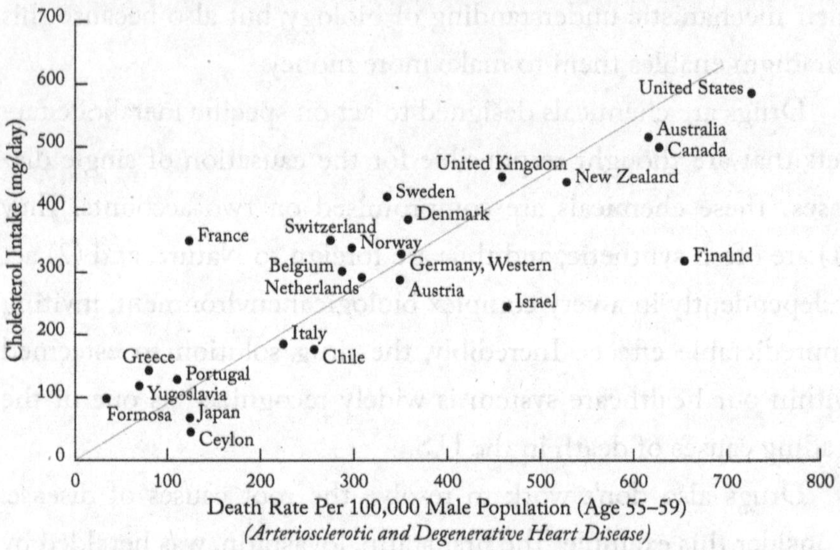

The graph refers to cholesterol intake, which increases with consumption of animal foods, and shows a wide range of heart disease mortality, low to high, for approximately twenty countries. The study suggests a theoretically achievable decrease in heart disease mortality of about 90 percent simply by eating a diet devoid of animal protein.

Another example of the bill of goods sold to us by the pharma industry are the drugs used for treatment of type 2 diabetes. This market in 2023 was $36 billion and expected to grow to about $70 billion over the next seven years.[28] Enormous amounts of money are being made, but at best these drugs merely mask symptoms, doing nothing to address the root cause of type 2 diabetes, which is food.

My son, Nelson, wrote and directed a film called *From Food to Freedom* (which he'll discuss later) wherein he showed how people with type 2 diabetes could heal within ten days through a whole food, plant-based diet. As discussed in that film, and now well-known by clinicians who are using a whole food, plant-based diet to treat this disease, someone with type 2 diabetes can sometimes improve so quickly with proper nutrition that insulin medication not appropriately adjusted can become toxic, even causing life-threatening hypoglycemia in the most severe cases. When a person taking these drugs suddenly shifts to a whole food, plant-based diet, they must be carefully monitored by a physician so these medications can be reduced in real time if necessary.

Considering this, drug companies should be required to inform any person with type 2 diabetes who is taking insulin of its potential toxicity should they begin treating their disease using a plant-based diet. Maybe these medications should come with the warning:

> *If you begin consuming a whole food, plant-based diet, your health is likely to improve. As your health improves, your body may produce more insulin on its own, requiring a reduction of your insulin medication. Taking this medication at your previously prescribed doses could lead to adverse health consequences, including hypoglycemia. For this reason, anyone with type 2 diabetes beginning a whole food, plant-based diet should be monitored by a physician.*

I believe future generations will look back at our overreliance on drugs as primitive. Today we know that our country's first president, George Washington, was likely killed by the three physicians who attended to him after he developed a sore throat and breathing difficulty. They administered a number of "treatments" thought effective at that time, including significant bloodletting. They literally bled him to death. Just as we look at those types of ideas today as medically barbaric, I believe people in the future will look at our overreliance on drugs as barbaric, because they produce limited benefits—mostly masking symptoms—while producing side effects that make drug treatment one of the leading causes of death in our modern society.

Despite these problems, we still love our pills. Indeed, we love them so much that we have even reduced food to pill form, which has given rise to the nutritional supplement industry. Supplements are generally single chemicals, removed from food, that act independently within a complex biological environment, resulting in uncertain and sometimes adverse effects. This is not the kind of nutrition provided by food, which involves thousands of nutrients and their analogs working together simultaneously and in unison.

Beta-carotene is a good example of this distinction. In 1981, a peer-reviewed paper summarized the results of a study showing that increasing beta-carotene intake (from food) decreased lung cancer among heavy smokers. But in a follow-up study, in which the researchers tested beta-carotene as a nutrient supplement prescribed to smokers, it showed an opposite effect, so much so that the study had to be terminated.[29] Beta carotene in food worked to decrease lung cancer risk, reflecting a whole-food effect, while beta carotene in a concentrated pill form had the opposite effect of increasing risk.

This should not be surprising to anyone who understands the wholistic nature of biology. We often think of nutrition as an arithmetic sum of specific functions of individual nutrients, but this is not Nature. There are thousands of nutrients in food, they work together in countless combinations, and they exist in seemingly countless plants and their parts—leaves, stems, seeds, fruits, and roots. The more we search for the properties of individual nutrients, and how and where in the body they work, the greater complexity (and confusion) we find.

It is true that single chemicals, whether drugs or supplements, can sometimes have limited benefits. But they are likely also to be short-lived and/or to produce unwanted side effects. Indeed, in most cases, the idea that we can control our health by consuming a single chemical working in isolation and at a level unnatural to the human body is silly. The real solutions can be found in food and the body's incredible capacity for self-healing.

After all these years of research and reflection, the wholistic nature of biology now seems so obvious. I often wonder, though, why it was not more apparent to me when I started my career, and why it is still not apparent to so many other people today. Why is such a seemingly obvious truth so hard for us to see? Let's examine this question, because I think the answer may pay dividends for us later in this book. Indeed, knowing why we see the world so blindly is essential if we are to fix anything about our world.

CHAPTER 4

ROOTS OF REDUCTIONISM

What drives reductionist thinking? Why, when there is a whole forest in front of us, the soil beneath, the sky above, and all the life throughout, do we examine and consider only the tree in front of us?

This is the question I have ended my career with. I feel satisfaction from the research we performed and the results it produced, but I am left with this foundational question about the way we see and don't see the world around us. And I suspect the answer to this question may also have implications far beyond my field of study.

I would like to share my thoughts on this question, and specifically what I call the roots of reductionist thinking, of which there are many. And I'll start with the one that is easiest to explain: laziness. We fixate on the one tree in front of us because it's intellectually the easiest thing for us to do.

A second root of reductionism is the quest for knowledge and reputation. We sometimes fixate on the one tree so that we can become a reputable expert. We stare at the cracks and crevices of the bark, the shapes and sizes of the limbs, the color of the leaves and their respiration rates, and on and on. We think if we can learn as much as possible about this tree, we can become an expert qualified to share our ideas with others. I performed plenty of reductionist research over the years, primarily in the laboratory looking at single mechanisms of action. As I shared earlier, this research was a valuable part of my career, and I do not mean to disparage other people who go deep into a particular topic in the same way. I respect such a quest for greater knowledge—as long as we do so with humility and don't claim knowledge we do not have, while always doing our best to see our narrow field of study from the perspective of the larger whole of which it is a part.

A third root of reductionism is bias—a topic I know well due to my experiences overcoming my own. I'll spend a little more time on this problem of bias because it can be so hard to see and challenge.

I've always liked the concept, long used in science, that, although we own opinions, we don't own facts. This means we should follow the formal rules of science to generate information that is then subjected to rigorous professional review. When we request research funds, design experimental studies, and interpret and publish results, we should do our best to control our biases.

There are three main types of bias in science: societal, personal, and institutional. By societal bias, I mean prevailing views that become fixed in our minds. These beliefs are generally called paradigms. They are almost always the result of reductionist thinking and often difficult to challenge by those not adequately trained in science and unfamiliar with the historical origin of these paradigms.

Despite the all-encompassing impacts of paradigms, we can't always blame them for our slanted way of seeing the world. We also have blurred vision because of personal bias. We all suffer from this handicap, although some of us suffer more than others. Personal bias occurs when we look at the world around us and see what we have been conditioned to see by our unique life experience, or what we want to see in pursuit of some kind of self-gain.

My personal bias came from my early life experience on a dairy farm, which instilled in me the belief that animal protein is the cornerstone of human health. I suppose you could say there was a prevailing societal paradigm at play here as well, but I know that all those hours milking cows, doing work that I thought had a meaningful purpose, played a major role.

A second kind of personal bias, one that has nothing to do with life conditioning, is the bias that comes from a quest for reputation, money, fame, and/or whatever else we might be seeking. This is a more purposeful kind of personal bias, where we see what we want to see for self-gain. There is, unfortunately, a great deal of this kind of personal bias in our world.

Finally, there is institutional bias. This is simply bias that is caused and reinforced by all I shared above and embedded in our institutions. There is tremendous bias, for example, in the bodies of government responsible for promulgating and enforcing public policy, as you'll see in the stories I'll tell.

There is a similar institutional bias embedded in our media. During the COVID pandemic, for example, the media repeated over and over, "Follow the science," when, to my knowledge, I never saw an individual "on air" truly trained in nutritional science, or who understood biology in a wholistic way. Through these voices, they did a poor job educating the public during the pandemic, who received plenty of education about masks, social distancing, and vaccines, but nothing about nutrition and its ability to strengthen the immune system. I'll dig into this further in a later chapter, when I tell the most important story from the pandemic.

And finally, institutional bias has also infected the world of academia. Historically, academic institutions have protected free speech through the tenure process, and as a result, were places where uncomfortable ideas could be pursued and shared. Within academia during more recent decades, however, I have noticed a disturbing trend. Rather than being spaces for free discourse motivated by a search for truth, academic institutions began institutionalizing within their walls many of the prevailing biases.

Tenure at the level of a university is not a mere guarantee of employment, as some may think. A tenured faculty position must be earned, based on merit. I was fortunate to have earned tenure quite early in my career. Had I not gained the protection afforded by tenure, it is likely I never could have written *The China Study* with my son, Tom, which included provocative and disruptive findings on the adverse health effects of animal protein. I know from life experience that the value of tenure cannot be overstated.

But recently, I learned of an ominous survey on tenure published by the American Association of University Professors (AAUP) for the period of 1980–2010. According to the AAUP, the proportion of faculty during that thirty-year survey period earning tenure in academia declined from 70 percent in 1980 to 30 percent in 2010, while the proportion of medical research funding provided by corporate sources—especially food and drug companies—increased, like a mirror image, from 30 percent to nearly 70 percent.[30]

Without the protection of tenure, and with funding provided by corporate interests, it's not hard to understand why so much academic research is conflicted, confusing, and misleading. As I have discussed, the dominant research paradigm is based on a reductionist perspective, which enables researchers to cherry-pick the small biological slice they want to study while blinding themselves to the complexity of the larger system surrounding that narrow subject of study. This approach enables researchers to design studies in ways that produce the outcomes sought by their corporate funding sources.

It's not surprising that we now see so many falsehoods perpetrated in places that used to stand for truth. Here are some that have found a home on so many of our college campuses:

1. Cancer is mainly a genetic disease.
2. Reliance on targeted drug therapy is safe and biologically defensible.
3. Saturated fat is "bad" and unsaturated fat is "good."
4. Environmental chemicals, not diet, are the main cause of human cancer.
5. U.S. dietary guidelines, produced by the U.S. Department of Agriculture, promote human health.
6. Dietary cholesterol causes heart disease.
7. Animal proteins are "high-quality" compared to plant proteins.
8. Nutrition was irrelevant as a preventive measure during the COVID pandemic.

I could write a dissertation on each of these, but for the purpose of illustrating a larger point, let's examine just one of these statements. As explained in my books, *Whole* and *The Future of Nutrition,* consumption of animal protein is more impactful than dietary saturated fat or dietary cholesterol in increasing serum cholesterol, the principal indicator of cardiovascular disease risk.[31] Yet this important information is still not widely known—possibly because the livestock and pharma industries have a vested interest in ignoring the connection between animal protein consumption, serum cholesterol, and heart disease risk. Livestock companies make money from foods high in animal protein, and drug companies make a killing on statin drugs used to reduce serum cholesterol.

We can understand biology more fully only from a wholistic perspective, but the roots of reductionism nonetheless run very deep to this day. These roots support a system that has produced untold amounts of disease and premature death—and trillions of dollars in wealth.

Given the immense amount of suffering and money involved, it's easy to fixate on the system that produces these outcomes. Indeed, in talking about our health crisis, we often talk of "the system," or "the industry," or "the government." I understand the value of this kind of macro analysis, but in this book, I want to dig a little deeper. I have lived my life challenging the status quo, and whenever I have done this, I have always encountered individuals who are making decisions that enable themselves and their institutions to benefit from the suffering of others. It's individuals who sustain immoral systems—sometimes because of ignorance, but often because of their own immorality.

In the next part of this book, I will share stories that show corruption in action—lifting the veil, so to speak—so we can understand what we are up against. The problem we face in building a plant-based world will not be solved through only macro-level solutions because it's a problem rooted in individual behavior. To share the truth of our health, we will require a more strategic and grassroots approach that can't be corrupted by powerful individuals vested in the status quo—a topic I'll leave for Nelson in the last section.

PART 2

LIFTING THE VEIL

CHAPTER 5

THE TRUTH BROKERS

Over my career, I have often felt exhilaration upon learning something I did not know before. I lived for these moments, each one of which reinforced my love for and commitment to the truth. Conversely, when I saw other people trying to suppress those truths, I truly felt like I had wandered into a very dark place. As I mentioned in the prior chapter, I will now share some of these experiences so that we can collectively understand the forces that have conspired to keep us sick, and from this understanding see our way to a better place.

I'll start with a story from the early 1980s—an exciting time for me and my research team. We performed our groundbreaking study showing we could turn the cancer process on and off merely by toggling between plant and animal protein. I also met my friend Dr. Junshi Chen and began developing plans for our research in China.

My research career seemed to be reaching heights I had never imagined, and as it did, doors opened for my participation in public policy. Around this time, I learned that a few highly influential members of my own professional nutrition society, the American Institute of Nutrition (AIN) of the Federation of American Societies for Experimental Biology (FASEB), were organizing a national committee to develop a nationwide "supreme court" of nutrition. Its formal name was the "Public Nutrition Information Committee." It sounded intriguing, because the public was experiencing a stream of questionable food, nutrition, and health claims in the marketplace.[32]

As the FASEB liaison officer on congressional matters, I was invited to join the committee. My assignment to this new committee, however, was as a nonvoting observer, and my role was simply to provide input and oversight. I carried with me an unbounded enthusiasm for the role of nutrition in human health and a now-lengthy history of research into the connection of diet to disease.

In our first meeting, the committee's purpose was explained in a proposed press release handwritten by the chair, Professor Tom Jukes of UC Berkeley, who had acquired a research reputation in pharmacology. The press release was intended to set the stage for the formation of this new nutrition information committee and it mentioned a few trendy yet questionable health claims to illustrate the confusion among the public on important medical and health issues.

I personally knew a few of the eighteen committee members, and in fact, one of them, MIT professor Alfred Harper, had written the principal reference for my faculty position at Virginia Tech in 1965. I was somewhat puzzled, however, by his membership on this new committee because his professorship had been endowed by the General Foods Corporation.

After that first meeting, conversations among the committee members continued by phone and by mail. A second in-person meeting was held a year later, at which the committee members planned to recommend permanent status for this committee within AIN. At the meeting, all eighteen committee members voiced support for permanent status, and Chairman Jukes asked for any further comments.

Although I was technically only an observer, I took this opportunity to speak. I had started seeing troubling industry connections and bias among committee members, so I spoke up to express my doubts as to whether this committee could be an impartial broker of nutrition and health claims. They were making it quite clear, at least to me, that they hoped to become a high-profile agency with concentrated power, and perhaps even with government regulatory authority. They were beginning to remind me of a McCarthyist tactic of the 1950s, which was still fresh on the minds of many. I was starting to see that, rather than protecting truth, this group wanted the power to do the opposite.

Chairman Jukes quickly became irritated with my comments. He hustled across the conference room and shook the arm of my chair, demanding I step outside the room, while the other committee members—who also appeared to be opposed to my comments—looked on. It was quite a scene. Just as I refused Jukes's request to leave the room, the meeting was abruptly interrupted by the pre-planned appearance of an Associated Press reporter at the door. He had come to pick up the news release announcing the committee's approval of permanent status. I realized with surprise that I had interrupted a planned celebration of accomplishment! At this moment, it became clear to me that the committee's year-long effort was simply a large-scale effort to present the public with a view of nutrition that served the interests of its industrial supporters.

I knew that a new president of AIN, under whose umbrella this committee resided, was taking office the day after this meeting. I sent the new president, Professor Doris Calloway of UC Berkeley, an overnight airmail to let her know that, although she might hear the committee had finalized a vote to recommend extending its work permanently, this vote had not in fact been completed. Calloway followed up a few days later, and to my shock, she dismissed the entire committee. Once she started naming replacements, she also appointed me as its new chair! I could never have imagined this turn of events. I used my new position to disband the committee altogether.

To say that the powerful people behind the proposed committee were angry would be an understatement. In fact, two of the main principals involved in the committee made arrangements to have me expelled from AIN—the first time an expulsion of a member had been attempted in its sixty-five-year history.

I was requested to attend an official hearing in Washington about my proposed expulsion. After the hearing, the group assembled to adjudicate the matter voted 6–2 against my expulsion. Though I was not formally notified of the specific charges against me, an ex-member of AIN's board, Professor Sam Tove of North Carolina State University, told me privately afterward that my opponents at AIN believed I had "betrayed" the professional nutrition community by disrupting their carefully laid plans. It had become clear to me that hostility toward me and my research was beginning to grow.

I was learning a big lesson on power and corruption, seeing how nutritional "truths" are engineered at the highest levels.

CHAPTER 6

MEDIA MEGAPHONES

The days of the famed and trusted news anchor, Walter Cronkite, seem distant, if we remember them at all. Many people today distrust the voices in our mainstream media, especially if we sense partisan political bias that is opposite our own viewpoints.

When most people think of bias in the media, they most often think of partisan political bias, but there is another kind of bias I would like to discuss in my next story, which is the kind driven by people and organizations with a financial agenda. Some of this is overt, such as the influence of drug companies who slather their commercial pitches all over our TV programming. Most of it is harder to see, though just as pernicious. I am speaking now of the supposed "experts" hired by media companies to communicate "facts" to the public, who are usually aligned with powerful commercial interests and whose testimonies reflect both personal biases and the prevailing paradigms of the day. I've witnessed this firsthand, as I'll share in a moment, but first I'd like to add a caveat.

I have done more than five decades of interviews with professional media on biomedical research and human health, mostly on diet and nutrition. Based on these experiences, I have come to believe many journalists try to act as responsible individuals and that the loss of public trust in the broadcast media should not be blamed on journalists alone (although there are many biased reporters). The problem of media bias, especially regarding nutrition, is generally more complex.

There is inherent difficulty in communicating biomedical science to nonexperts, as it often relies on unpronounceable words and unfamiliar concepts. Researchers struggle to communicate using familiar language, and journalists are not trained in fundamental science or in medical science. It's two sides of the same coin; we struggle together to reach each other and to communicate and understand discoveries of complicated concepts.

Further exacerbating this problem is the fact that most new discoveries in biology and human health, often referred to as "science" by the media, are becoming more detailed and isolated from a natural context (in a word, reductionist), often leading to flawed and conflicting results. This is especially true for nutrition, which is best understood by considering its all-important biological complexity and comprehensiveness. Explaining nutrition in reference to isolated and independent events and substances is very much the language of business and enables soundbite news reports that oftentimes do not reflect what is true.

With this caveat, I would now like to share a story of a particularly egregious example of media bias. This story starts in 2016, when I received a phone call from the legendary British

Broadcasting Corporation (BBC), a worldwide radio and TV news network I had long admired. The BBC invited me to do an interview for a television program and offered to come to the U.S. for the interview, when they would also interview my friend and colleague Dr. Caldwell Esselstyn in Cleveland.

Coincidentally, I was passing through the Cleveland area at that time, so we ended up doing the interview near Cleveland, riding in a golf cart around an apple orchard. The TV crew stood on a platform mounted to the front-end of the cart, with their BBC colleagues in London listening in live and asking a few questions as well. Unfortunately, this event did not go as I thought.

Though I expected an honest exchange, I quickly started to sense an agenda. My interviewer was a University of Cambridge scientist, Dr. Giles Yeo, who announced to me at the outset that he was an unchangeable carnivore. He also made clear that he was a geneticist, suggesting to me that he believed disease to be genetically determined. And in fact, Yeo was quite argumentative, showing that our interests were very different.

Though I remained cautiously optimistic about the film after our day of interviews, my hopes were unfounded. Unfortunately, the subsequent BBC film bore little similarity to what I had envisioned. Through creative editing, it denigrated the effect of nutrition on human health and disease, and it aligned me with other interviewees who had earned reputations for making questionable claims that led to legal proceedings. It seemed to me that Yeo wanted to associate me and my ideas of nutrition and disease with quackery.

Efforts were made to discredit our research in the laboratory, our human study in China, and Esselstyn's heart disease reversal study, all of which showed the power of nutrition to promote human health without pharmacological intervention. That same day, for example, Yeo had also interviewed Esselstyn and three of his patients who had reversed serious heart disease through a plant-based diet, but those testimonies were omitted from the film. These patients told impressive accounts of their own health experiences. Were they too convincing?

Yeo also downplayed the book I authored with my son Tom, *The China Study*, which, at that time, had already sold more than two million copies and had been translated into well over forty foreign languages. He seemed to want me to retract my research about the ability of whole food, plant-based nutrition to prevent and even reverse experimental cancer in the lab, whose findings were later supported by our human study in China.[33] He also wanted me to avoid mentioning that there was virtually no obesity in rural China, which I had noted in *The China Study*.

Yeo seemed to have no interest in hearing my thoughts on obesity, including that calorie intake per unit body weight in China at the time of the study (in the early 1980s) was 20 percent *higher* than in the U.S., calling into question the common idea that obesity is a direct result of caloric intake. Indeed, similar observations have been made in the nutrition literature for at least a hundred years. But none of that made it into the film.

Yeo believed that he knew how to conduct research studies and interpret results better than I did, and he warned me to be careful about spreading what he clearly perceived as misinformation about people's health. I reminded him that I usually consider multiple and varied studies to form my views—but that comment was omitted from the film, leaving the impression that I relied only on our human observational study in China. He asked whether I understood the limitation of using the correlations recorded in this type of study to draw conclusions; of course I understood. I knew this limitation well, and I took care in our study design to eliminate the problem by taking a wholistic approach, incorporating hundreds of variables to find larger patterns reflective of the diet-disease connection. To be frank, the fact that he even asked this question suggested to me that he had little—if any—understanding of the design of our research study.

Like so many others in our community, Yeo seemed to have a reductionist view of biology, which I confirmed through later research. I was not familiar with him or his research interests before the interview, but I have learned since then that he has professionally authored with colleagues a relatively large number of peer-reviewed manuscripts on obesity. His professional work suggests a preference for the idea that obesity has a genetic cause, as well as for the use of pharmaceuticals to treat this supposed "disease." This approach has proved quite profitable.

I recall during the 1980s and 1990s considerable conversation on the increasing rates of obesity and a need to solve the problem. There was lots of interest in doing research on the topic, but getting support required that obesity be considered a distinct disease and thus made a target for profitable interventions. Increasing conversations occurred from 1990 until 2013, when the American Medical Association recognized obesity as an independent disease requiring treatment, a requisite first step for marketing anti-obesity drugs. Now, several such drugs are on the market, with the usual side effects and questionable lifetime benefits. Not surprisingly, market size is now expected to grow to over $100 billion by 2030.[34]

After viewing the finished film, Dr. Esselstyn and I wrote to the BBC expressing our concerns. Though they acknowledged receipt of our emails, we did not hear when or if a response might be forthcoming. Because the BBC had already released their documentary in the U.K. and because it was rumored that it would soon air in the U.S. on CNN, I posted my email to the BBC on the Center for Nutrition Studies website, as did Dr. Esselstyn.[35]

Here is the eventual response from the BBC director's office after he received my initial complaint and saw what I posted to our website:

> *The BBC is disappointed to discover that Prof Colin Campbell has chosen to publish a critical article about the BBC programme Horizon: Clean Eating – The Dirty Truth, transmitted on the 19th January 2017 without any comment from the programme makers.*
>
> *The BBC rejects the accusations of conflict of interest and agenda that Prof Campbell makes about both the programme and its presenter Dr Giles Yeo. As with all BBC programmes this film was edited in line with BBC Guidelines, and accurately reflects the views of Prof Campbell that he expressed in the interview he conducted with the BBC.*
>
> *The film has been critically acclaimed and has prompted an important national debate about dieting advice and evidence based science.*

And here is my reply (also posted on our website):

> *Mr. Liddell says it all, as he did in his response to Dr. Esselstyn's commentary, when he states that the "film makes frequent reference to the importance of eating a healthy diet." This means nothing to viewers who are known to have a wide variety of ideas as to what is a healthy diet. In fact, the documentary clearly infers that material of myself and Dr. Esselstyn may not be healthy.*
>
> *I fully understand that only portions of an interview are used in a final documentary, but the choice of material is what gives meaning to the final product. BBC severely slanted the story, basically denigrating the role of nutrition in maintaining health and treating disease in favor of Dr. Yeo's belief that human health is best understood and practiced by determining genetic pathways that pave the way for drug development. It is this latter practice of ignoring nutrition that demeans what human health really means—in favor of a pharmaceutical strategy that is hugely expensive and far too often is without merit. I am confident that it creates wealth for the few at the expense of health for the many.*

As I noted, Yeo favors research that leads to the development of drugs to address obesity. And in his publications, he has stated that his research has been supported by the U.K.'s Medical Research Council and the Helmholtz Alliance Imaging and Curing Environmental Metabolic Diseases, which is an alliance of thirty research teams and research centers "enhanced by cooperative alliances with Sanofi Aventis Pharmaceuticals."[36] The fact that the BBC conveniently ignores this admission raises some questions about their views.

To add insult to injury, I learned later that Yeo was claiming in lectures that he succeeded in getting me to say that I did not have evidence to support the conclusions and hypotheses on human health and disease that I reported in *The China Study*. Of course this was not true. Though Yeo's false claim has been challenged, for example by Klaus Mitchell, a young man in the U.K. who runs an online plant-based news program, it nonetheless circulated around Europe.[37]

My experience with the BBC, a global broadcast news organization comprised of dozens of foreign news bureaus and thousands of journalists around the world, was both surprising and disappointing. I have long respected its reputation as a non-commercially funded media source, its fair-minded (I thought) reporting of the news, and its unparalleled broadcast reach. Unfortunately, while the BBC's funding is public, the BBC still seems to be waving the pharma banner high, supported by "expert" voices like Giles Yeo.

Like academia, the media *should* be protectors of the truth. But too often they are the perpetrators of lies that help to line the pockets of people and industries happy to trade away the health and happiness of others for their own gain. This story does not end with the media, though, because there is another player in this game that's just as important—government.

CHAPTER 7

A FORBIDDEN TOPIC IN GOVERNMENT

It's easy to pick on governments these days; their failures are so obvious. In the U.S., the federal government has managed to run up (as of the time of writing) over $38 trillion in debt. Much of this spending supports a healthcare system that makes and keeps people sick, while also imposing a massive cost on the government through Medicaid, Medicare, and other health-related initiatives (the largest category of spending in the U.S. federal budget).

Government can serve as a safeguard to protect the public from predatory forces, but history shows the reverse to be true more often than not. To be clear, I am not a virulent anti-government person; indeed, I feel there is a role for government in our society and acknowledge the good outcomes government sometimes produces. However, I also see this as more the exception than the rule, a trend we can see clearly in our public health policies.

The story I will tell to make this point is about a door that opened for sharing new nutritional information in the 1970s that was slammed shut by powerful interests, setting the table for decades of corruption that strengthened the influence of people and organizations profiting from our sickness.

During the late 1970s, our experimental research results on nutrition and cancer were getting more attention and causing more provocation. It was around this time that I decided to take a one-year sabbatical leave from Cornell to work in the office headquarters of the Federation of American Societies for Experimental Biology (FASEB) in Bethesda, Maryland, near the NIH campus. At that time, FASEB was a federation of six nationally prominent biomedical research societies (a total of about 80,000 professionals).

Among a variety of responsibilities during my sabbatical, I served on several professional committees at the national level, including expert committees associated with the National Academy of Sciences (NAS), considered the premier U.S. science academy. NAS is a private nonprofit organization that establishes committees of science experts to evaluate biomedical topics of national interest, with most projects being funded by the government.

Two of the most significant NAS committees involved preparing reports on diet, nutrition, and cancer: one on evaluating the scientific evidence for the effect of nutrition on cancer (1982), and a second recommending research priorities (1983). The first committee, of which I was a part, produced a report that was the most sought-after report in the history of NAS at that time, drawing considerable public and political attention. Along with two other committee members, I was invited to give testimony on our committee's findings to U.S. House and Senate committees.

The 1982 NAS committee on diet, nutrition, and cancer arose in response to an earlier 1976 U.S. Senate committee chaired by Senator George McGovern on diet and health.[38] This committee recommended lower consumption of meat, which drew a sharp rebuke from the livestock industry. This persuaded the politically sensitive committee to publish a second, slightly revised 1977 edition that suggested less red meat but allowed "white" meat (chicken and fish),[39] although even this slightly modified report prompted industry complaints that it was too political and not sufficiently based on the "science."

The earlier McGovern report also prompted discussion about the funding focus of the government's National Cancer Institute (NCI). Senate hearings were held, and the director of NCI, Arthur Upton, was invited to testify on how much NCI funding was dedicated to research on the connection of nutrition to cancer. I was surprised to hear him testify that only 2–3 percent of his huge cancer research budget was dedicated to that investigation.

I later spoke with McGovern on these matters when he spent some time at Cornell. He said that, although he considered his dietary goals project within the U.S. Senate to be the favorite work of his career, it came with a price tag. As many as six of his Senate colleagues from farm states lost their re-election bids in the subsequent 1980 elections, likely because of the industry's response to this report. Promoting anything concerning diet, however modest it may be, can be politically costly.

Now, back to my story about our NAS committee. With that foundation laid by McGovern in the 1970s, our subsequent 1982 NAS committee revisited the role of food in health. We focused, as was the reductionist custom, on the effects of specific nutrients on cancer, with individual chapters on calories, lipids, carbohydrates, dietary fiber, vitamins, minerals, and various other non-nutrient dietary constituents. Initially, no chapter was intended for protein, so I suggested to the committee that this omission should be addressed. Although my colleagues were not particularly interested, they nonetheless allowed me to write a draft chapter on protein that our full committee would review. Some of them expressed concern that drawing attention to protein, especially of animal origin, might create a political debate too hot to handle. Even so, I wrote it, and the chapter was included.

Despite the inclusion of my chapter, the final report still showed sensitivity to the interests of the industry. In particular, it highlighted total and saturated fat as the strongest diet-related cancer risk, intentionally making saturated fat the "bogeyman" to diverted public attention away from the truly harmful animal protein–laden foods that bring along the saturated fat (which, unlike the protein, can be modified to produce "low-fat" dairy and other animal products). As it turns out, saturated fat is largely biochemically inert in the body, yet it's still villainized, even among many influential voices in the plant-based nutrition community.

Despite attempts by the committee to show sensitivity to industry interests, the animal livestock industry was quite displeased when our final report was released in 1982. The report quickly became newsworthy, drawing considerable public and political attention. It suggested an incremental shift in diet in the direction of plant-based foods, and the chapter I authored noted increased risks associated with overconsuming animal protein.

The food industry did not like the direction the report seemed to be taking and were ready with their own publication: a booklet summarizing the comments of fifty-six well-known scientists and other thinkers who almost all sharply questioned the report's conclusions, which was placed on the desk of every U.S. representative and senator as soon as the NAS report was released.

Following the release of our 1982 NAS report, I was invited to testify to several congressional and other governmental health committees. Along with fifteen to twenty TV interviews and multiple print media news releases, these opportunities created more public exposure for me than I had expected or needed at that time. I was especially sensitive to this exposure when I was asked about our research findings on protein and cancer. The warnings my colleagues had given me early in my career about conducting research on animal protein and cancer hung over me like a dark cloud. But our research findings, however controversial, were becoming more convincing. I had no choice but to continue following the data and sharing it with the public.

I came to Washington full of naïve optimism and energy but left doubting the capacity of nutritional science to filter through government in a way that could serve the public interest. Because of the influence of powerful corporate interests, speaking about the harmful effects of animal protein, in particular, seems sacrilegious in government.

CHAPTER 8

EAT MEAT, DRINK MILK. LOTS OF IT!

Moving forward to the turn of the century, our next story is again focused on how government communicates to the public, this time on how it formulates its dietary guidelines. Before jumping into this story, I need to provide a little background on the process used to formulate the U.S. dietary guidelines.

The U.S. dietary guidelines we now observe were first issued in 1980 by two highly reputable individuals I knew well, Dr. Mark Hegsted of Harvard and Dr. Alan Forbes, Director of Nutrition at the FDA. Both men spent five to six months writing a draft of the dietary guidelines in 1978 and 1979. The U.S. dietary guidelines report that they produced has been updated every five years since then.

Not surprisingly, corporate interests have been active in manipulating these guidelines throughout the past forty-five years. In 1999, the well-regarded Physicians Committee for Responsible

Medicine (PCRM) filed a lawsuit against the U.S. Department of Agriculture (USDA) and Department of Health and Human Services (HHS) for claims arising out of the appointment of the Dietary Guidelines Advisory Committee for the year 2000. The lawsuit alleged that the USDA and HHS had catered to the interests of the food industry by not disclosing conflicts of interest among committee members. While the PCRM contended that six of the eleven committee members maintained "inappropriate ties" to the dairy, meat, and egg industries, the court addressed the USDA's omission of information about an industry payment of over $10,000 to one committee member. The court acknowledged that such connections could render "a Committee member . . . financially beholden to a person or entity that had an interest in how the [Guidelines] might be amended."

And this is not the only evidence that the guidelines' positions on animal products can be manipulated. For example, one way to keep the unusually alarming evidence about animal protein out of the public eye is to recommend "upper safe levels," or ULs, of nutrient intake for all nutrients so as not to focus only on protein, and then set a UL for protein intake high enough to satisfy political and economic interests.

To get a little more into the weeds on the policy-making process, the U.S. dietary guidelines are developed by a joint committee of scientists of the U.S. Department of Agriculture and the U.S. Department of Health and Human Services. This joint committee develops guidelines based on recommended intakes of individual nutrients determined by the Food and Nutrition Board (FNB) of the Institute of Medicine (IOM, now called the National

Academy of Medicine), whose reports also are updated every five years. Together, these two groups (FNB/IOM and USDA/HHS), essentially working together, are meant to provide scientific judgment, authenticity, and governmental authority.

Let's take the 2000 U.S. dietary guidelines again as an example. The news release for the report read, "To meet the body's daily energy and nutritional needs while minimizing risk for chronic disease, adults should get 45 percent to 65 percent of their calories from carbohydrates, 20 percent to 35 percent from fat and 10 percent to 35 percent from protein."[40] The statement that caught my attention, of course, was the UL for protein: 35 percent of total diet calories, which remains the UL as of this writing. This metric was preposterous—if Americans were to consume protein at this level, it would lead to serious health consequences. In my opinion, this protein UL is the most consequential decision in the history of nutrition public policy.

I'll explain. Illustrated in the chart on the following page is a minimum daily requirement for protein of about 5–6 percent of calories, and a "recommended daily allowance" of about 9–10 percent of calories, to make sure that everyone consumes enough protein. However, the U.S. population consumes much more than the recommended amount, an average of 17–18 percent and a range of 11–22 percent of calories, represented by the horizontal black line in the chart. This average, although higher than the recommended level of protein intake, would not be a health problem if we were eating protein provided by plant foods like legumes, nuts, and seeds. But in reality, we don't do that. Instead, an average of 75 percent of our total protein intake comes from animal foods that kill us prematurely through heart disease, cancer, and other degenerative diseases.

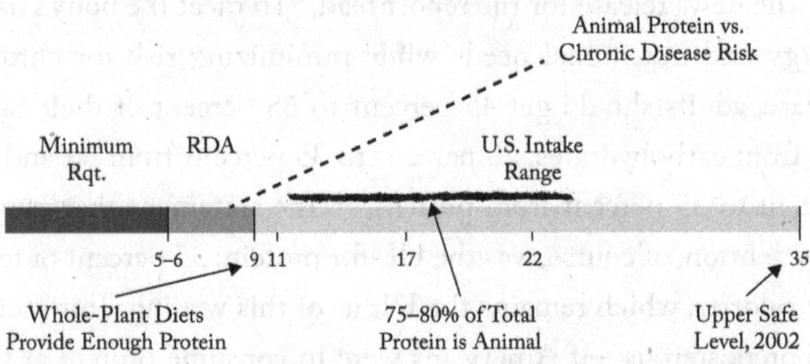

The sloped dashed line in this chart collectively represents the data I shared in the first part of this book, namely that increased consumption of animal protein increases disease occurrence. Remarkably, this association of protein with increasing disease risk begins within the range of our usual consumption of animal protein and suggests why such a high percentage of us die prematurely. How, then, can it be possible that the committee decided that dietary protein is "safe" up to an incredible 35 percent of total diet calories, a percentage nearly *four times higher* than the recommended allowance for protein? Without doubt, this new UL of 35 percent for protein was irrational, irresponsible, and unmindful of evidence. Whose interests did the committee have in mind when they formulated the new guidelines?

I personally knew a member of the FNB/IOM committee

responsible for making the nutrient recommendations used in formulating the dietary guidelines, a longtime friend and colleague who directed the toxicology program at the FDA, and so I phoned him to ask why the committee made such a surprising recommendation. Though a congenial fellow, he reacted somewhat defensively to my questions because he had spent the last six months coordinating the IOM committee's discussion of fat, carbohydrate, and protein requirements, including the UL recommendations. But after some discussion, he offered, "Colin, you know that nutrition is not my area of expertise." In fact, he was merely filling in for the lead discussant on the topic, Professor Vernon Young at MIT, who had spent much of his research career investigating the amino acid composition of animal protein, which supposedly gave it its "high quality." Young had left the committee before its last meeting to serve on the board of directors of Nestlé Corporation.

I had known Young ever since he came to MIT in the 1960s as a young professor, just a few months before I left for my faculty position at Virginia Tech. He and I started our careers in similar ways. As a professor at MIT, he became involved in childhood nutrition programs, especially in Central America, while I did much the same thing in the Philippines. Both of us had initially accepted a popular opinion in our research community that childhood malnutrition would be well served not only by assuring adequate calorie intake, but also by assuring adequate protein intake, especially protein of "high quality"—as in, animal-based protein.

Subsequently, however, our research careers veered in strikingly different directions. Young mainly conducted highly technical (reductionist) research to explain the "high-quality" designation of animal protein, among other characteristics, as a function of its amino acid composition. In contrast, I became interested in investigating why animal protein promotes experimental cancer, whereas plant protein does not.

Young, an immigrant from Wales, was personable, and sincerely devoted to his views on protein. On one occasion, when I was presenting our research findings to a large audience at an international nutrition conference in Seoul, South Korea, he exclaimed from the audience during the Q&A period, "Colin, don't take animal protein away from us; it's good food!" That comment, not unlike those made by many others in our profession, suggested that the perspectives on animal protein taken by many of my colleagues in nutritional science are influenced as much or more by personal preference than scientific evidence.

About a decade later after I'd last seen him, Young was appointed to the IOM committee described here and led its discussion on this new UL set for the macronutrients fat, protein, and carbohydrate. In the committee's 1,100-plus page report, a large section was devoted to his own research findings on the amino acid composition and supposed "high quality" of animal protein.

But he wasn't the only one in the committee working to encourage protein consumption. In my experience as a member of similar committees hosted by IOM, it was customary for staff management to review (and edit) a final copy of the report, then to prepare a news release for the public. A member of the IOM committee at that time, who helped prepare the news release, was Professor Cutberto Garza, who directed my nutrition department at Cornell University.

Garza also chaired the companion USDA/HHS committee that uses IOM reports to develop dietary guidelines. As noted earlier, the committee's work was later challenged by a PCRM lawsuit, although the suit was filed against the USDA and HHS, not the committee members, for failing to disclose potential conflicts of interest of its members.

In his role on the IOM committee, Garza participated in the development of an upper safe limit (UL) for protein intake—a figure that some experts, including myself, questioned as insufficiently supported. The UL was then cited in the IOM's news release and subsequently informed the USDA's dietary guidelines, which Garza also helped oversee. To my knowledge, it was unusual for one individual to play such roles in both the IOM and USDA committees.

Garza and Young had enjoyed a years-long friendship, which began when Garza earned a PhD from MIT, working with Young. I first met Garza when I returned from a year-long sabbatical at Oxford University, where I had worked with my colleagues on the China study. In my absence, Garza was recruited to Cornell to be the director of our Division of Nutritional Sciences, my academic home. Though I did not know Garza, I looked forward to my meeting with him upon my return.

In that meeting, he acted as if we had previously met and, although cordial, he seemed somewhat uneasy. I was puzzled. Many years later, I learned why—Garza had been present at the Public Nutrition Information Committee, whose establishment I had successfully blocked (see the "Supreme Court of Nutrition" story shared in chapter five).

Talking shop in that meeting, he told me that in his early experimental research as a pediatrician he had acquired data suggesting that infants did not need as much protein as generally recommended, prompting my asking where he published this. He replied, "Oh, no, I didn't publish that," leaving me to wonder why he failed to do so.

It was Garza, Young, and others of their colleagues who had established the 35 percent protein UL, which not only allows, but encourages, consumption of animal protein-based foods. Why not eat all we can get! I can hardly imagine a more effective marketing scheme for animal-based food products than to be supported by official government recommendations, at the expense of the very public they claimed to serve.

For those of you who may question whether this guideline has had an impact, I'll tell a quick story from the world of dietetics. The professional community of dietetics, especially the influential 112,000-member Academy of Nutrition and Dietetics (AND), is a very influential communicator of dietary information to the public.[41] I was invited three times to present keynote seminars to their national conventions over a period of three decades, usually sponsored by their Vegetarian Practice group. However, on the most recent occasion, when I registered, I was given a welcome bag that highlighted their "AND Partners": GlaxoSmithKline, Coca-Cola, PepsiCo, National Dairy Council, Danone, Nestlé, Mars, and Aramark, which is a major service provider of school lunches. In my presentation, I shared our research findings and spoke with candor . . . and have not been invited back again.

I also have presented to state-level AND conventions, most

recently in Newark, New Jersey, which afforded me the opportunity to inquire of the three hundred attendees their awareness of the UL for animal protein. I began by asking how many knew the general concept of "upper safe limits" for nutrient intakes. Only about five to ten people raised their hands. I then asked whether they knew that the UL for protein was 35 percent of total calorie consumption. Most hands went up. Clearly, these dieticians, presumably a representative sample of the profession as a whole, accepted the idea that we can consume essentially all the animal protein we want—far above the recommended level of 9–10 percent of total diet calories.

I've long personally admired dietitians and public health nutritionists and the work they do. I've taught many of these people at Cornell in a nutritional biochemistry course required for nutrition majors and premed students, among others. Many of these students later obtain their registered dietician credentials, then work closely with clients and patients, especially in clinical and hospital settings. In my experience, they constitute the most direct linkage that the public has to the topic of nutrition.

I find it offensive that dedicated professionals must rely on nutritional guidance developed within a system heavily influenced by industry funding—including support from the dairy, livestock, processed food, and pharmaceutical sectors. I consider the promulgation of the 2000 dietary guidelines to be *one of the most harmful developments in American health policy*. An official upper safe level of 35 percent protein, excessively rich in animal-sourced protein, is unconscionable.

CHAPTER 9

CONTROLLING PURSE STRINGS

As my last two stories demonstrated, government can communicate falsehoods, but it also can use the power of the purse to suppress truth, which is the point of my next story. This is a story of a tremendous lost opportunity to forever change our understanding of disease, specifically how it forms and then progresses. This story took place in the late 1980s and features as the villain the Department of Health and Human Services—a powerful voice for human health and home to the National Institutes of Health that fund biomedical research.

NIH has long been the premier biomedical research funding agency in the U.S., and perhaps worldwide. I have been involved with NIH in multiple capacities since 1969. At that earlier time, NIH was composed of twenty-seven individual institutes, each dedicated to a specific disease or health discipline. My research program was continuously and handsomely funded for three decades by one of those institutes—the National Cancer Institute (NCI).

In 1988, when our China project had been well underway for five years, I joined my colleague, the famed epidemiologist Sir

Richard Peto of the University of Oxford, in applying for NIH funding for a potentially groundbreaking research project. It was a follow-up to our existing project in China, whose data were then being tabulated for publication by Jill Boreham, Peto's colleague at Oxford. This new, much-expanded project would have collected data from 500,000 subjects, a significant increase from the 6,500 subjects in the first study. Each of these subjects would provide blood samples every two years, which we would freeze and store for later analysis when subjects received a disease diagnosis. We also decided we would maintain a constant number of subjects by adding new participants to the study as older participants passed.

A liquid nitrogen freezer, automated for sample storage and retrieval, was already established by colleagues of mine in Europe. They were organizing their European Prospective Investigation in Cancer (EPIC) study, but without having the repeat collections every two years that we were proposing. I went, with my lab director, Dr. Martin Root, to visit their operation in Lyon, France. It was impressive, located in a small building with minus 80°C freezers.

The first China study had focused on sixty-five counties, comparing average death rates and possible causes for about fifty diseases. This study had already been hailed as world-class research, but the proposed extension would be even more remarkable. Due to the number of subjects, the number of measured biomarkers and their combinations, the study's duration, and its experimental design, it would have provided us with much more information than the previous study. Instead of comparing sixty-five county averages at one point in time, it would compare 500,000 *individuals* at recurring collection times until each subject's death, perhaps spanning a study period of thirty years or more—meaning it would be 2.5 to 5 million times more informative. By measuring a large array of biomarkers prior to disease onset and during disease progression up until death, we could then see patterns reflecting each point in this process. We could identify the subsets of biomarkers at each point in time that were most predictive of the onset or progression of a particular disease. We also would have had an opportunity to revisit stored samples to measure newly discovered biomarkers found to be predictive of a particular disease process.

Best of all? This revolutionary study would have had an unparalleled potential to evaluate once and for all our dependence on drugs. We could have seen more clearly the development and progression of disease, and therefore the impacts of nutrition on a disease process in real time. We could have seen disease reversal unequivocally, captured in changing biomarker patterns. This would have challenged the current dogma that assumes diseases are the outcome of specific mechanisms of action that are best controlled using drugs. Nutrition could have been shown to be far more effective, with those impacts measurable in a precise way that had never before been possible.

When our proposal for this groundbreaking study was submitted to NIH, we had already established strong professional and personal relationships with our Chinese colleagues, who were positioned at the highest professional and governmental levels and who had demonstrated an outstanding data collection and recording system, possibly the best in the world. Chinese government officials assured me that they were able and willing to participate in this new project were it to proceed.

After its submission to NIH, the proposal was reviewed and received a high "fundability" score, with approval granted for an initial $7 million in funding (equivalent to around $20 million today). I was asked to come to the NIH headquarters in Bethesda, Maryland, to accept this funding in person because the program involved the National Heart, Lung, and Blood Institute in addition to the National Cancer Institute. My assistant, Dr. Thierry Brun from the Pasteur Institute in Paris, joined me. The NCI representative, Dr. Dan Nixon, who I personally knew, placed a $700,000 "down payment" check on the table as soon as Dr. Brun and I entered the meeting room.

Strangely, a representative of a third NIH Institute was also present: Dr. Otto Bessey from the National Institute on Aging. His institute was not party to the funding of this project, but he had been invited by NIH authorities to share a letter sent to his office and signed, "Friends of the Institute on Aging," which said that our study in China, which was not yet published but which had then been running for five years, was fraudulent!

In response to that letter, the Heart Institute director said he could not move forward with our new project until this strange outside claim was addressed. It seemed to me, however, that this claim was impossible to respond to because it was *signed anonymously*. This was unprofessional behavior at its worst. In my opinion, it was clearly the work of the same power syndicate I had seen elsewhere.

I was told at the meeting that the anonymous letter would be "taken under advisement," and I would hear more within two weeks. But over the next two weeks, I heard nothing. It seemed clear to me that powerful people behind the scenes had influenced the NIH to derail our newly approved nutrition project, despite its professionally determined scientific merit and its approval by the grant review committee. This reflected an act of exceptional corruption and dishonesty. I had never heard of a project that had been approved within the NIH grant review system not being funded.

I knew exactly how unusual this was. I had lots of experience with this system and knew well the overall structure and administration of the NIH research funding program, from both sides of the table. I already had substantial experience reviewing requests for funding of other scientists as a member of several so-called "study sections." I had an unbroken record for over three decades requesting and successfully receiving funding for our research in the laboratory and in real-life settings (human study projects in China and the Philippines), as well as working with varied cancer research organizations using the same review system. In the Division of Nutritional Sciences at Cornell, a nutrition department ranked first in the country, my program was the best funded and the most professionally published, supporting a research and training program with the most graduate students in the department.

I have no doubt that this professionally approved but never funded project would have offered an unparalleled opportunity to investigate an exceptionally complex and time-dependent biological effect of nutrition on health. Undoubtedly, this would have been a game changer, with promise unlike any biomedical health research project ever done before or since. It could have exposed, at a fundamental scientific level, how seriously compromised is our pharma-driven medical system.

Some time later, and perhaps not coincidentally, I came to realize that the National Heart, Lung, and Blood Institute was deeply immersed at that time in supporting the pharmaceutical industry's launch of the first statin, a medication that lowers cholesterol. In retrospect, I am certain that we would have acquired evidence that diet is far more effective in controlling and even reversing heart disease than a statin.

Regrettably, I did not pursue this $7 million funding award any further. I considered reporting what had happened to the media, but I didn't follow through, for a couple of reasons. First, I was experiencing a difficult bureaucratic challenge with the 1989 expansion of the first China study, which would have included the same 130 villages surveyed in 1983 plus another eight new villages in rural China and thirty-two new localities in Taiwan. A particularly stubborn administrator in the NIH financial office was blocking the release of an already approved $300,000 grant to my Taiwanese colleagues because, according to our campus business office, he "had fought against those commies in Korea" and had no interest in sending the check. I asked our university provost to personally accompany me to the NIH, if needed, to secure the money for this new proposal, and I let the NIH know that if we were not successful, I would take our story to *The Washington Post*. Thankfully, the check was then released.

Concurrent with all this, I had become overburdened with other responsibilities and with an extremely difficult personal health challenge arising from my earlier work with dioxin at MIT, which I discussed earlier.

The loss of funding for a research project that would have

revolutionized healthcare in our world was most regrettable. After this experience, it was clearer than ever that my earlier views of government as a force for good were naïve. I saw firsthand, once again, that powerful people use government as a tool to further their own financial interests, even when they cause immense suffering among the people whom the government is supposed to represent—and who foot its bill.

CHAPTER 10

SETTING THE STAGE FOR A PANDEMIC

As my previous stories have shown, there are powerful systems in place that reinforce, and sometimes encourage, our illness. We get sicker and sicker, and the interests who push this process along get richer and more powerful. Our sickness often takes the form of chronic degenerative disease, and it was this kind of illness that set the stage for the worst infectious disease crisis in the modern era.

When it became apparent we were headed into the COVID pandemic in the spring of 2020, I rather naïvely thought perhaps our government and the media might finally be forced to share the truth of nutrition's importance to our health, especially when we learned early in the pandemic that well over 90 percent, perhaps over 95 percent, of the people dying from COVID had one or more medical conditions, *mostly lifestyle-related chronic conditions.*[42] Thus, it stood to reason that plant-based nutrition, which can address these conditions, could provide a powerful defense against the virus. This information, however, was not shared with the public. Indeed, the website of the Centers for Disease Control and Prevention failed to even mention the word "nutrition" in its advice for protection against the worst effects of a COVID infection.

I knew of the connection of nutrition to immunity not only from my general understanding of nutrition and its connection to chronic diseases, but also because of certain data from our research in China. Early in the pandemic, I went back to our original data and found remarkable evidence for the protection plant-based nutrition can provide against a viral infection.

When our project in China (conducted in 1983) was repeated in 1989, we collected data on a serious virus, hepatitis B (HBV), which is the principal cause of liver cancer, the third leading cause of cancer deaths in the world. As in the original study, we collected a large amount of information on nutrition, some socioeconomic data, and many disease death rates, including for primary liver cancer caused by HBV. Our subjects' average consumption of animal protein, as a percent of calories, was only one-tenth of U.S. consumption. Nonetheless, individuals consuming animal protein at the higher end of these lower levels were less likely to form HBV antibodies when testing positive for the active virus (antigen) and were more likely to die of liver cancer. Conversely, individuals consuming more plant-based foods were more likely to form antibodies and were less likely to die of liver cancer.

This evidence showing the protective effect of plants on the immune system was based on eleven highly significant correlations, with only one unexplained correlation, all pointing to the same effect. The results were some of the most convincing and highly significant research findings we ever obtained. They also confirmed findings on animal protein and virus-induced liver cancer in experimental animals that we ran in the early 1990s.[43]

This is the same nutrition that prevents and reverses chronic degenerative conditions like type 2 diabetes and heart disease, so this kind of benefit for the immune system is not surprising. To think otherwise would be the same as thinking that what is good for one arm is not necessarily good for the other arm. Our body is one fully integrated system.

After I reviewed the HBV data from our China study, I wanted to share this important information with the public. The pandemic had wreaked havoc around the world, and the information that would have saved the most lives was unknown to almost everyone. Governments and the mainstream media were failing us.

If they weren't going to share this information, I thought I could at least prepare a manuscript on our HBV findings. My manuscript was titled "A Nutritional Link for COVID-19?"[44] This might have sounded strange to people steeped in the pharmacologic paradigm, but sometimes the truth seems strange, especially when you are sitting inside a box with limited visibility of the world outside.

I submitted this manuscript to two distinguished medical journals where I had previously published. To my surprise, I received an administrative response from both that I had never seen before in my professional career: the journals' staffs refused to send the manuscript for professional review.

I am no stranger to the publication process. I have served on journal management and editorial committees, have published well over three hundred papers, most being peer-reviewed, and have reviewed the manuscripts of many others. Manuscripts submitted for publication are *always* sent for professional review, provided the manuscript is on topic. Merit is determined by scientists, not by administrative staff, who are more likely to bend to organizational interests. I was shocked that the staff members refused to send out my manuscript for review, especially given my long history of reputable research, my publication history, and the urgency of this topic.

I do believe that both journals were seeking research evidence that might control this new coronavirus infection, but I suspect the people behind them were steeped in the official government mantra focused on shutdowns, social distancing, handwashing, masking, and, most of all, vaccines. My claim about the effects of nutrition must have sounded strange indeed. I also suspect that there could have been, let's say, market sensitivities guiding their decision-making. Nobody wants to get on the bad side of pharma, do they?

This is an example from the science media, but of course, the mainstream mass media also turned a blind eye to the importance of nutrition in the pandemic. In fact, I happen to personally know certain prominent media personalities, and I can confirm that they understood the connection of nutrition to immunity. One of them even read my book and claimed publicly that it had influenced his own life, yet I never heard him on any of the evening news shows discuss the power of plant-based nutrition to protect against the worst effects of a COVID infection.

Rather than share positive and empowering information about the importance of nutrition, the media stoked fear among the public throughout the pandemic. Then, on the backside, they framed vaccines as our only way out. Newscasters were constantly urging the public to follow official guidelines, as voiced by "experts" who were merely following the science, but who rarely encouraged anyone to eat healthier or to be careful with their diets!

In making this argument, by the way, I should be clear that I am not an anti-vaxxer. Vaccines saved the lives of vulnerable people who were at serious risk from a COVID infection. I am not saying that a vaccine was not warranted, but I am a scientist, not affiliated with pharma or any of their stooges in government and the media. This means I have a more nuanced view, driven by real science, not the "science" often touted by the media.

My wife and I, both in our eighties when the vaccine was released, stood by our beliefs on this issue and made the decision not to vaccinate. Yes, we were older, but we both were healthy because we were consuming a healthy diet and exercising outdoors regularly. Ultimately, we tested positive for the virus about fifteen months later and experienced nothing more than the usual symptoms of a flu-like response for about a week.

The decisions made by the administrators of the two major medical journals not to submit my virus research manuscript for professional review, at a time when the virus vaccines were making big news, showcased the systemic bias of the media, governments, and—of course—industry. Our research evidence was an inconvenient truth at a sensitive time, and more easily suppressed than shared. Our evidence showing a likely safe way to protect against the most severe consequences of infection would have been a threat to the incredibly lucrative COVID-19 vaccine business.

The results for the pharma companies that sold the vaccines were nothing short of stunning. Since late 2020 the four Western firms that have dominated the COVID-19 vaccine market—Pfizer-BioNTech, Moderna, Johnson & Johnson, and AstraZeneca—have together rung up about **$195 billion in vaccine revenue.**[45] The windfall peaked in the first two full-launch years: **sales of about $63 billion in 2021** and **$61 billion in 2022.**[46] Compare this to the *entire* global vaccine market in the last pre-pandemic year of 2019, which generated sales of about **$33 billion**, and to Pfizer's previous blockbuster, the pneumococcal shot Prevnar, which topped out at **$6 billion a year.**[47]

It also goes without saying that the FDA has facilitated the business of the pharma industry for years. For instance, the FDA commissioner from 2017 to 2019 was Dr. Scott Gottlieb, who then departed the FDA to sit on the Pfizer Board of Directors. I have no proof of any quid-pro-quo arrangement in this case and so I will leave this for your judgment.

Again, I am not opposed to vaccines per se; I know there were vulnerable groups during the pandemic who benefited from the vaccines. What I am saying is this: we should not have sold vaccines as the *only* way out of the pandemic, and we should not have limited the recommended preventive measures to social distancing, handwashing, masking, etc. We should have also shared the empowering truth about nutrition and its' connection to immunity.

CHAPTER 11

CHILDREN FOR MONEY, PART 1

There are two more stories to share before I finish. As we've seen over and over in these chapters, people in positions of power frequently trade away the well-being of the public to further their own interests. Tragically, this includes the health of the most vulnerable among us—our beautiful children.

I'll never forget Antonia Demas, one of the kindest souls my wife and I ever had the pleasure of knowing. Sadly, Antonia passed away a short while ago, but her memory lives on in the people whose lives she touched. Antonia was a graduate student at Cornell University who wanted to use her high-level training to foster societal change, in this case for the health of our young children.

For those who may not know, graduate studies involve students earning advanced degrees (MS, PhD, and other professional degrees) while conducting original research studies under the oversight of a committee of three to four faculty, who eventually assess their work to provide final approval. I served on many such committees, serving as chair of more than forty for my own graduate students. But Antonia's was especially notable. I served on her committee as her nutrition advisor.

For her research, Antonia undertook a project to investigate a more effective way to teach healthy nutrition to elementary school children. She was personally and profoundly devoted to children, having raised two of her own. She had also worked as a volunteer and consultant in related fields for twenty-five years, first in the Head Start program in her children's school in Vermont, where she experimented with developing strategies to educate very young children about eating healthy foods. Her approach utilized hands-on experience and sensory-based education. She discovered that when children are engaged in a developmentally appropriate manner, they enjoy trying and eating these new foods.

In 1997, her family moved to upstate New York, where she volunteered in her children's classrooms. She also used the experience she gained in Vermont to begin working as a part-time consultant in other schools, teaching kids in an afterschool program to cook nutritious foods using healthy USDA commodity foods—basic government-funded foods provided to schools. She was very aware of the political aspects of the USDA-administered National School Lunch Program and was careful in designing and describing her project in a way that would not provoke too much public attention or unwarranted criticism. She simply wanted to test ways to educate children in a manner that would provide lifetime benefits. It was her love for children that she wore on her sleeve.

Unfortunately, she heard many comments denigrating her work, so she decided to earn a master's degree and then a PhD from

Cornell University in order to gain more professional credibility. Her master's degree was in professional studies and human service, nutrition education, and human development. For her PhD, she envisioned a project that I thought had tremendous potential, not only as a subject for her thesis, but also for her post-graduation work.

For many years, she had been volunteering in the elementary school in her small hometown of Trumansburg, New York. It was about ten miles from the Cornell campus, which provided an opportunity to do her PhD graduate work there as well. She made a detailed research plan for a year-long research project, obtained permission from the key players at the school, and trained volunteers to collect data.

Antonia educated the children on food and nutrition as well as including them in food preparation. She also provided them with opportunities to work in a small demonstration garden, to compost food using leftover scraps, and to work with the school food service. She even invited Cornell students from other countries to discuss their home countries' food practices with the school children. Without doubt, any observer of this project would agree that Antonia had an unusual gift for teaching young children, and her experience-focused approach proved effective in winning kids over to a healthy way of eating.

But she didn't stop at teaching kids. She also invited parents for evening discussions and engaged the local food store and farmers in providing the right kinds of food to buy. She worked hard to engage the whole community.

In the meanwhile, Antonia collected data on her observations and experiences for her doctoral dissertation. It became clear that both the children and their parents were unusually enthusiastic and supportive. At the conclusion of her graduate studies, her dissertation received two national awards, one from the USDA and the other from the Society for Nutrition Education. Based on all she learned, Antonia subsequently published a children's book on food, nutrition, culture, and the environment, *Food is Elementary*, which also became the name of her curriculum program.

Antonia's accomplishments also attracted the attention of the Under Secretary for Food, Nutrition, and Consumer Services, Ellen Haas, the Clinton-administration official charged with overhauling the National School Lunch Program's menus. Haas told Antonia that she was the only person who seemed to have a solution for getting kids to want to eat healthier food, as her dissertation data showed. Haas then invited Antonia to apply for a demonstration grant to embed her *Food is Elementary* curriculum in several school districts as a proof-of-concept.

It's at this point that her story takes a sinister turn. Having heard about her success, a senior faculty member at Cornell persuaded Antonia to share her dissertation and draft grant proposal, saying he would shepherd her paperwork through the Cornell nutrition department. Shockingly, when the package reached the USDA, her name had been replaced with his on the principal-investigator line, while leaving her curriculum, data, and narrative unchanged. Inexplicably, Cornell later concluded that although the failure to properly credit Antonia was "insensitive and careless," it did not constitute plagiarism or academic misconduct.

The grant was approved, but a few years later, Haas resigned from her position and the USDA's priorities shifted. After her departure, the agency cited "non-performance" of the project, then headed by the Cornell professor who had taken it from Antonia, and canceled the grant, directing Cornell to return the unused funds. In its review, the USDA noted that the team heading the project had not initiated the promised classroom pilots, collected no usable student data, and failed to collaborate with participating schools. What they didn't mention was any possible institutional bias inside the USDA against the new approach Antonia had developed—a cause Ms. Haas had courageously joined.

Antonia saw her project stolen and then canceled, but she was not one to give up. She went on to launch the Food Studies Institute in rural upstate New York. Working with her daughter Ariel, she continued her pioneering work in various schools, quietly proving that children will choose the right foods when given the right education.

The theft and cancellation of Antonia's project and the resignation of her champion inside the USDA represented an immense lost opportunity to change a program that drives profits at the expense of our children. The National School Lunch Program feeds kids with animal-based and processed foods that can cause allergies, digestive problems, obesity, and chronic diseases later in life. The federal government foots the bill for much of this food, and nearly 70 percent of the commercial value of these USDA-supplied foods, and over 40 percent of their poundage, are animal products. Every reimbursable school meal must include a meat or meat-alternate—though almost always a meat is provided—and a full eight-ounce serving of milk. To make matters worse, commodity animal products are routinely converted into nuggets, breaded patties, deli meats, and pizza, which are often high in sodium and full of other unhealthy ingredients.

The question is this: why do we feed such unhealthy foods to our kids? Is it because we are stupid? Do policymakers think chicken nuggets are healthy? Perhaps there are a few people like this, but the primary answer once again is simple: money. The foods purchased by the federal government for the National School Lunch Program are purchased with the goal of supporting farms—especially farms producing animal-based foods. In other words, we are trading the health of our kids to support the animal agricultural industry.

As I've shared, I grew up on a farm, although our farm bore little resemblance to the ones supplying the school lunch program today. These corporate farms are far larger than the farm I grew up on. Nevertheless, there are many people in farming communities today who retain the values we held dear on our farm. I loved farm life and our neighbors who farmed, and that is why I think farm families are owed the truth about our health as well. Powerful forces can suppress a truth like this for only so long, and once people understand this truth, they will begin to demand a selection of food that our farmers today are not supplying. When that time comes, producers will need to understand these important truths every bit as much as consumers do.

Indeed, truth is the most important element of a truly free market. It's what enables producers to meet the genuine needs of consumers. The USDA is supporting an agricultural system that is destined to change. Rather than facilitating this change in a way that would enable producers to adjust over a reasonable period of time, our government seems hell-bent on perpetuating a lie that will eventually harm farmers who don't have the capital and time to make the changes that will be required—changes that are destined to happen because of the courage and vision of people like Antonia Demas.

Though Antonia has passed, her light still shines on through her daughter, who is now running Antonia's Food Studies Institute (foodstudies.org),[48] and through everyone else fighting for the same truth to be told.

My story about what we do to our young people in the name of money does not end here. I will share one last story, this time about the people in our society who are the most powerless of all—our babies.

CHAPTER 12

CHILDREN FOR MONEY, PART 2

Nutrition is important throughout our lives, starting even before birth. Mother Nature met our need for good quality nutrition by providing mothers with breasts to nourish us at the moment we first see light. Indeed, there may be no better representation of living in alignment with Nature than a mother breastfeeding her child. We live in a society disconnected from the natural world and the principles that define it, but even in our modern materialistic society, the superiority of breastfeeding is now well accepted, even advocated, among professionals. Advisories by prominent health authorities are consistent, recommending at least six months of breastfeeding, and preferably another six to twelve months while blended solid foods are introduced.

As this book has shown, however, an obvious truth can easily be ignored and even purposefully suppressed. But to suppress an obvious truth at the expense of our babies is unforgivable. There are many great philosophers and spiritual teachers who have spoken of the beauty of a new life, innocent and full of curiosity and wonder for the world. How could we possibly do anything to hurt a baby? Let me explain.

I should begin by acknowledging that, despite Nature's gift of breast milk, there has always been an understandable need for artificial infant formula—some mothers die in childbirth, and others produce too little milk, are parents to adopted babies, or cannot nurse for other reasons. When breastfeeding is not an option for a parent of a newborn, the most important thing is to make sure the baby is fed as healthfully as possible. Though breast milk is better than infant formula, infant formula is most certainly better than nothing.

Because of situations like these, people have been creating breast milk replacements for many years. One of the most famous people to do this was the German chemist Justus von Liebig. Often referred to as the father of organic chemistry, agricultural chemistry, and scientific materialism (reductionism), he created an artificial infant formula in the 1860s and soon began selling it.[49]

The infant formula industry developed further in the twentieth century, and by the 1970s it had become quite powerful—and also the center of a scandal. Nestlé Corporation, a Swiss company and one of the largest companies worldwide to produce and sell infant formula, had come under fire for engaging in overly aggressive marketing tactics. It was estimated that such tactics—led by companies such as Nestlé—contributed to roughly 212,000 infant deaths every year in low- and middle-income countries, triggering an international boycott as well as many protests and lawsuits.[50]

I did not know much about infant formula or the misdeeds of companies like Nestlé, but in the 1990s I found myself with a front row seat to a story unfolding at Cornell involving the infant formula industry. This story concerns someone I introduced in chapter eight due to his involvement in the increase of the government's "safe" upper limit for protein consumption—Cutberto Garza, director of Cornell's Division of Nutritional Sciences.

A pediatrician by training, Garza worked closely with Professor Vernon Young at MIT, who chaired his PhD committee (also introduced in chapter eight). In 2001, Young joined the board of the Nestlé Corporation while continuing his faculty role at MIT and, like Garza, became deeply involved in national and international policymaking committees.

A third character in this story was the late Professor Michael Latham, physician and world-renowned director of the international nutrition program within Cornell's Division of Nutritional Sciences. Latham was perhaps the best-known scientist in the world focused on childhood malnutrition, especially in tropical and semitropical countries, and one of his primary interests was the use and abuse of infant formula and commercial formula manufacturers' exploitation of women in developing countries. In fact, my return to Cornell in 1975 was prompted in part by a presentation he had delivered at Virginia Tech regarding his and my common interest in childhood malnutrition. Over two hundred students, mostly from outside the U.S., studied in Latham's program at Cornell, and several were later named to senior positions in their home countries' health programs.

In 1997, at the International Congress of Nutrition (ICN) in Montreal, Canada, a sharp dispute arose over ICN's acceptance of funding from infant formula, food, and pharmaceutical multinationals. This dispute gained steam following a keynote presentation by UNICEF deputy executive director Stephen Lewis. Speaking to an audience of 1,500 nutritional scientists from around the world, he criticized ICN organizers for accepting funds from infant formula companies like Nestlé. His presentation was met with a standing ovation.

Latham was surely one of the people who participated in the applause, and later that day, he and some of his colleagues held a press conference on the topic of infant formula. He was well known for his expertise on this topic and had even provided testimony to congressional committees on multiple occasions.[51] Although Latham agreed that breast milk substitutes should be available in all countries, he argued they should primarily be used for infants whose mothers could not provide breast milk. He adamantly opposed the promotion of infant formula in any form because he knew that the use of infant formula had resulted in *millions* of infant deaths due to weakened immunity, mixing with contaminated water, dilution and malnutrition due to the product cost, and other factors—a fact recognized by no less than the World Health Organization (WHO). It should also be noted that he was not alone—an overwhelming number of professional nutrition scientists were deeply concerned with the promotion of these products by Nestlé and other infant formula companies, who had been subject to harsh criticism for decades due to highly predatory and misleading marketing.

Upon returning home from the conference, Latham, still disappointed by the funding sponsorship of ICN by Nestlé and others, then learned that a press release he had prepared in Montreal and sent to the Cornell News Service (the news and media relations office at the university) had not been issued; Latham claimed Garza blocked its release. Latham, one of the most gracious and accomplished people I have ever known, wrote to Garza:[52]

I was very disappointed on my return to Ithaca from the Congress in Montreal to learn that your actions had resulted in my press release not being issued by the Cornell News Service. This appears to me to be an infringement of academic freedom. This, of course, is a serious matter. I now need to consider what steps to take, because I strongly believe that faculty members have a right to publicize their views even if these are not ones shared by superiors, or even by the university.

You very well know my views on the issue of financial support and promotion by the infant formula industry. We have discussed this openly and have agreed to disagree. But I will not be silenced. I believe that it is unethical and creates a conflict of interest when the infant formula industry provides financial support to individuals who, or conferences which, are addressing problems related to infant nutrition or health. This is because the promotion of infant formula reduces breastfeeding, and can result in harm to infants. Few doubt that the promotion of breastmilk substitutes in this way has resulted in infant disease, malnutrition and deaths. I realize that although you are a proponent of breastfeeding, you disagree with me about funding and that you willingly accept money from the baby food industry.

This Memo then is a plea to give me and the rest of your faculty freedom to express opinions, and have the Cornell News

Service release these views even if they are not your views and may annoy corporations from which you seek financial support.

It is now more than twenty-five years later that I am reviewing some of this correspondence between the younger Garza and the older Latham, who as I noted has since passed. Latham was a key participant in the infant formula quarrel nationally and internationally at the highest administrative levels. In his congressional testimony, Latham admonished formula companies for their aggressive promotion of infant formula product over breastfeeding in developing countries. Although his testimony gained considerable media attention as well as promises from these companies to suspend those practices, I personally saw little evidence that those promises were ever kept. What I did see was the industry continuing to expand its power and influence even on my own college campus.

Research in academia on food, nutrition, drugs, and other health issues has become more and more beholden to corporate interests, a reality which has especially affected my younger colleagues. They must think carefully if they want to keep their positions. As I discussed in the opening to this book, institutional bias nowadays is highly responsive to external sources of funding, and academic freedom is almost gone.

This infant formula matter is yet another example showing how nutrition information, when shared with the public, can be manipulated for private interest. I once assumed, long ago, that infant health would always take precedence over corporate profits. How naïve!

Perhaps I should have known better, based on my own

experiences—including a bizarre episode involving one of our children. After the birth of my second son in a rural hospital in Virginia in 1967, a nurse came to my wife's room to give her a shot, not telling her why. When my wife asked, she was told the shot was "to dry you up." We were amazed that a hospital would do this as a matter of standard policy—without even telling, much less asking, the mother! Can I suspect that this reflected a successful marketing campaign to sell a product? Needless to say, my wife refused the shot, and breastfed all five of our children, each for at least eight and sometimes over twelve months.

But times are changing. As I mentioned at the outset, researchers and medical professionals are now acknowledging that breastfeeding is optimal for infant health. That's a step in the right direction, but we unfortunately still have a ways to go. The promotion of infant formula has not ended. During a UN-affiliated World Health Assembly meeting in 2018 in Geneva, Switzerland, a resolution was introduced to reaffirm that "mother's milk is healthiest for children and countries should strive to limit the inaccurate or misleading marketing of breast milk substitutes." However, as *The New York Times* reported, "the United States delegation, embracing the interests of infant formula manufacturers, upended the deliberations."[53]

The U.S. delegates, at this later-than-late date, not only opposed the resolution but also turned to strong-arm tactics. Ecuador, which had planned to introduce this resolution as a rather innocuous measure on behalf of developing countries, was informed that the U.S. "would unleash punishing trade measures and withdraw crucial military aid" if Ecuador followed through with the resolution. Even the dozen-plus assembly participants who agreed to speak with journalists about the meeting "requested anonymity because they feared retaliation from the United States."

Some American delegates also suggested that the U.S.—the largest funder of WHO—might cut WHO's financial support, as it was encouraging breastfeeding instead of infant formula, if they didn't get their way. An Ecuadorean official said she was shocked because she "didn't understand how such a small matter like breastfeeding could provoke such a dramatic response."

I do, because I have lived many years on this planet fighting for a just cause, and have witnessed how many people are willing to stand on the side of deceit and injustice to gain money and power. These last two stories about how we are willing to trade away the well-being of our children and infants for shallow self-gain shows the depth to which many people are willing to sink.

CHAPTER 13

MY CLOSING REFLECTIONS

It seems so simple, yet for our world today, this advice is profound:

The ideal diet is one comprised of whole plant foods: fruits, vegetables, grains, legumes, nuts, and seeds. These foods provide all the essential nutrients needed for optimal human health. Based on incontrovertible research evidence, at least 80–90 percent of the diseases responsible for premature deaths and shortened lifespans can be prevented and often reversed by eating a whole food, plant-based (WFPB) diet.

Refined oils from a bottle should not be used as an ingredient, but whole fats from plant-based foods like nuts, seeds, and avocados are important to our health. Salt and sugar can be consumed at low levels and need not be eliminated, because there is no research suggesting any benefit from the complete elimination of these elements, especially in the context of a whole food, plant-based diet. Having some flavor in your food is key to sustaining a healthy and joyful plant-based lifestyle.

This simple but vital truth, perhaps one of the most important of our time, has encountered hostility from individuals and institutions who see the world through biased lenses, as well as those who are pursuing their own gain at the expense of others, often behind closed doors.

The primary tool used by these powerful forces is reductionist science. Of course, great reductionist discoveries have been made, such as those leading to computers, jet planes, space travel, instant worldwide communications, video presentations, electric cars, and now artificial intelligence. I say "great," because materialistic inventions can be truly impactful and useful *if developed and used with the right intent.*

But the development of new technologies, which are exciting and sometimes even helpful, does not always benefit human health, which arises from a complex biological environment that cannot be described purely through linear thinking. Materialistic, linear tinkering in the realm of human health is naïve and can be dangerous. This tinkering is the rationale for pharmacology, for example, which primarily manages symptoms—often poorly—while also creating toxic side effects.

Nutrition is the opposite of pharmacology. When a person eats well, the nutrients in that food work together in the body to create health, as Nature intended. This working together of endless nutrients and nutrient-like substances in food is a concept that makes me think of the great artists from the Renaissance period, like Leonardo da Vinci and Michelangelo. Their art was not merely brushstrokes and colors. Instead, these virtuosos used their intellect to show our interconnectedness with the larger natural world.

Because of biology's near-infinite complexity, whose depths we have only begun to explore over the last century or so, nutrition can also be interpreted like art, music, poetry, and literature. This perspective on nutrition captures the same sense of wholeness that is reflected in those works, and in Nature itself.

When health research relies on materialistic reductionist thinking, as it so often does, it becomes more vulnerable to control, especially by those who are motivated by money and power. These people and organizations often achieve their ends by controlling the funding of new research investigations then controlling how findings of this research are reported. This has been our story in the modern era, especially in the U.S., which now has the lowest life expectancy among economically developed countries, the highest per capita use of pharmaceuticals, and the highest per capita healthcare costs. These facts reflect a healthcare system that is far less effective than most people realize.

If we go back in time, way back in time, a more wholistic view prevailed. I have often read that the ancient Greek philosopher and mathematician Pythagoras (570–490 BC) was a vegetarian. He greatly influenced Hippocrates, Plato, Socrates, Aristotle, and other Greek scholars who spoke about health in a more wholistic way, aligned with Nature.

Hippocrates, who is widely regarded as the father of medicine, famously said, "Let food be thy medicine and medicine be thy food." Ironically, medical schools across the U.S. charge their students upon graduation to adhere to the Hippocratic oath, which is generally understood as the duty to put patients' welfare first, avoid harm, respect confidentiality, and practice with compassion.

Hippocrates understood that at the core of all this was the use of food as medicine. Yet, our modern medical system ignores this message, and relies on drugs, often using them in ways that harm their patients.

As a science, nutrition is dismissed by the medical system in two exceptionally important ways. First, few American medical schools give any substantive education in nutrition, and whatever little nutrition they teach is often incorrect. Second, the premier medical research agency in the U.S., perhaps the world—the National Institutes of Health—refuses to add an Institute of Nutrition to their twenty-seven institutes and centers. I was professionally active in advocating for both changes, giving nutrition lectures at medical schools on behalf of the American Medical Association and organizing the development of a new nutrition and cancer "study section" with my NIH colleagues, which we hoped would lead to the establishment of an Institute of Nutrition. Both efforts failed. Certain leaders of the NIH, in particular, were adamant in their denial, as if the idea of nutrition is beneath medical science. I was told that it should be enough that they already had nutrition programs in some of their institutes. I found it very strange and offensive that they saw nutrition, the very foundation of human health, as at best a minor add-on to other medical specialties.

The way we currently examine and discuss public health is trending in a frightening direction. My earlier experiences living within a research community that uses formal, transparent procedures to search for truth, with efforts to control and minimize bias, may very well be history. I was fortunate for much of my career to have the freedom to seek and then speak truths without being confined to institutional rules or preferences—in other words, to have academic freedom. I learned its importance more with time, especially after so many efforts were made to deny it, such as through decreases in the tenured positions that enable scientists to report unpopular observations and conclusions. I doubt there is any scientific discipline more in need of tenure than in health science, where lives are at stake.

At the beginning of my research career, I was warned by my colleagues that questioning the "high-quality" nature of animal protein would be problematic. I ignored their cautionary words and zealously chased the truth. With hindsight, I can see that they were right, perhaps more so than they realized. But if I had the chance to pursue my career all over again, I would do the same thing. I had no choice; I had to follow my heart and do what I believed to be right.

Fortunately, I came along at a point in history when it was not so difficult to do the right thing. My proposals for research funding, based on sound scientific principles, were well received by professional scientists. Over the years, I successfully submitted thirteen detailed and rigorously evaluated proposals for research funding, in which I made plans to investigate the surprisingly harmful health effects of animal protein–based diets on human health. Though the NIH funding application process was highly competitive, all thirteen of these proposals were successfully funded, sustaining a continuous research period of about thirty years.

This was a time when funding, as well as professional publication, was awarded primarily based on the authenticity of the science, regardless of whether it might be contrary to public opinion.

Science was, at its root, a search for truth. That was then . . . but not now. We have lost much of our academic freedom, while funding for research and related projects is less public and more corporate. And when commercially funded research produces findings that fail to support commercial interests, funding is discontinued, public health be damned!

President John F. Kennedy understood this problem in the early 1960s when he announced a huge increase in public funding for biomedical research. This required a system to objectively evaluate requests for funding by qualified researchers. But following President Kennedy's assassination in 1963, the tide began to turn. During Lyndon B. Johnson's administration, interest in seeking research funding from the corporate sector started to grow. Now, half a century later, requests for government funding to do experimental research are much more controlled.

The stories I have shared in this book reflect the ever-growing influence of corporate interests in the academic and public policy worlds. They use reductionist perspectives and tools to create information they can label as "science," which they can then control and commercialize. They do not want science that seeks broad understanding of natural phenomena. They want research on specific causes, specific explanatory mechanisms, and specific treatments that can be commercially exploited, especially if they can be protected by intellectual property instruments. These forces then use unethical stooges to enforce their control over the rest of us.

I say all this not to make you feel hopeless, but because we first need to understand our reality and its problems in order to make things better. The good news is that there is always a solution to every problem.

Unfortunately, we have not been very good at finding systemic solutions in our own plant-based community. We have made some progress here and there, but our message is still considered a fringe idea. Our world remains awash in animal-based and processed foods. Our communities often have few healthy restaurants, our healthcare system continues to ignore plant-based nutrition as a tool for healing, our media continues spouting what I can only describe as nutritional nonsense, and our governments mostly just reinforce this corrupt and costly status quo. Even with all the resistance I faced over the course of my life, I never expected that the truth I and many others have fought for would still be largely hidden from public view all these years later. We have not been very good in our plant-based community at movement-building, and we need to understand why.

FROM DESPAIR TO HOPE

I would like to turn the rest of this book over to my son Nelson. As the eldest child in our family, he had a front row seat to my career. He witnessed many of my challenges and successes, which I shared freely with our family, often around a big circular table when we sat down together for dinner in the evening. Over the years, Nelson has developed a deep understanding of the science I spent my life pursuing. He also has developed quite a knack for effectively communicating this science.

Just recently, I attended a small nutrition immersion event hosted by Nelson and his wife, Kim, in the mountain town of Saluda, North Carolina. He asked Karen and me to sit in the audience as he delivered two seminars over successive days telling the story of my life and explaining my scientific research. I was so impressed by his understanding and his communication abilities, and I know beyond any doubt that he can successfully carry this torch well into the future.

Since he was a young child, Nelson also has had a strong social, environmental, and ultimately political impulse. We always thought he would get involved in elective politics, especially after he volunteered for our local congressional representative, even serving for a while as his deputy campaign manager while still a student at Cornell. He became dismayed by the political process, though, and he decided to make changes in other ways.

Nelson has been joined in this work by his wife, Kim, who also attended Cornell, studying education, child development, and nutrition. She then became a schoolteacher, but she retained a lifelong interest in all things culinary. She eventually became an expert on plant-based foods, writing three nationally regarded cookbooks, developing plant-based products for manufacture and retail sale, and engaging in culinary education. Together, Nelson and Kim have been working hard for more than fifteen years to advance the plant-based nutrition message.

For this final section of the book, I asked Nelson to share his ideas for accelerating our transition to a plant-based world. We know this world awaits us in the future. In Nelson's first film, *PlantPure Nation* (available on Amazon Prime and Apple TV),

Kentucky legislator Tom Riner cited an old proverb: "The truth is a stubborn thing . . . it doesn't go away." He is right, but timing is important. We need to accelerate the shift to a plant-based diet, not only for health reasons but for environmental ones, which Nelson will discuss further. His ideas are innovative, optimistic, and hopeful. And more than anything else, we need hope. It's hope that motivates the courage necessary for making difficult changes.

So many years and so many adventures have come between where I started and where I am now. But I am still that boy who grew up on a farm, immersed in Nature. I remember many things from those days, such as the lessons my parents taught me about hard work and integrity. However, more than anything else, I remember the beauty around me. I think our primary goal in this modern world should be to rediscover that beauty. If we can open our eyes to the majesty of Nature, both outside and inside ourselves, perhaps then we can begin to see things more wholistically, and if we do this, we can build the healthier and more sustainable world we envision.

PART 3

THE CHANGE

(NELSON CAMPBELL)

CHAPTER 14

LOSING AIR

I vividly remember the evening when our family sat down for dinner around a large round table and my father excitedly shared that he had discovered he could turn the cancer process on and off in the lab, like a light switch, merely by toggling between plant and animal protein. This was a heady time for him. On the heels of this breakthrough, he met his Chinese colleague Dr. Junshi Chen, and together they soon began conceptualizing a pioneering research study in China from the more wholistic perspective he had been developing—one that he felt would better capture the relationship between food and disease.

I remember being amazed at what my father had accomplished. I thought he would win the Nobel Prize in Medicine, and that it would only be a few years before the truth he discovered would begin transforming the world. The reality turned out to be far more challenging.

I watched my father take his message to Washington, D.C., only to be rebuffed again and again by powerful people benefiting from the status quo. I saw him struggle with severe health issues related to his dioxin research at MIT and with growing hostility in his community, including on the Cornell campus. He became severely stressed and increasingly dejected. Our family saw this and encouraged him to take matters into his own hands by writing a book to share with the public what he had learned. He also took charge of his health by undergoing unconventional medically supervised fasting therapy to leach the accumulated dioxin out of his system and by eating a stricter form of the whole food, plant-based diet with my mom.

These steps renewed my father, especially when he saw the almost immediate success of the book he wrote with my brother Tom, *The China Study*. It had been difficult finding a publisher because several of the large, mainstream publishers he had approached, who were supposed to understand their industry better than anyone, told him his book would never sell unless it was rewritten as a "diet" book, with plenty of recipes. My father ignored that myopic view and finally found Glenn Yeffeth, founder of BenBella Books. Unlike many of his colleagues in the business, Glenn was willing to take a chance on my father and his message—a decision that paid big dividends. *The China Study* quickly became an international bestseller, and it was this book, more than anything else, that launched the plant-based nutrition movement.

A few years later, Brian Wendel, John Corry, and Lee Fulkerson released the pioneering documentary film *Forks Over Knives*. This movie told the story of my father's research, while also telling of a clinical application of these ideas led by Dr. Caldwell Esselstyn, who had been trained as a surgeon and was recognized for his expertise on thyroid and parathyroid surgical procedures. My father first met Dr. Esselstyn in the 1980s when he learned that Dr. Esselstyn had put a group of heart disease patients (referred by cardiologists at the Cleveland Clinic) on a meat-free diet, although they were still consuming some dairy products. After meeting my father and learning of his research focused on animal protein and chronic disease, Dr. Esselstyn decided to take the patients in his study off dairy as well, fully shifting them to a plant-based diet. The ensuing dramatic results demonstrated that heart disease is often a reversible condition.

Forks Over Knives followed in the footsteps of *The China Study*, and blazed the way for other books, online media, and films, including our own feature documentary, *PlantPure Nation*. Millions more people learned of my father's research in the following years. It started to look like the plant-based nutrition message would, indeed, change the world, just as I had thought so many years ago sitting at our family's dinner table.

It was during this time that I made a major life commitment. I knew I was entering what I thought would be my most productive years and decided to join my father in sharing his health message. I had developed entrepreneurial skills that I believed could be useful, along with a vision for a grassroots strategy that could circumvent the established power structure. I had seen my father struggle over the years to share his message with policymakers in Washington, and I had lost all hope in the idea that government could function independently of the people profiting from our poor health.

As I embarked on this journey, my first task was to explore the best approach for motivating people to make a difficult life change. Changing what we eat, which is so foundational to our daily lives, is not easy. I was familiar with this challenge from many years of trying to explain my father's research in a way that might open the people around me to the possibility of change. So, I set about researching various potential approaches, and I soon found a program that had received considerable attention: the Complete Health Improvement Program (CHIP). This thirty-day intensive education program had achieved some success, producing an average reduction in total cholesterol of 11.3 percent.[54] I was impressed by this effort, but I did not think CHIP was scalable to large populations due to its intensive format, involving multiple educational sessions over several weeks. I also wondered if even better results could be obtained.

I started thinking about plans for a more scalable and experiential program, and I came up with the idea for a ten-day program that I initially called a "Jumpstart." This program included one or two educational sessions on the front end, plant-based meals prepared daily, a few other "touch points" along the way, and pre- and post-program biometric testing, including lipid panels and sometimes fasting glucose tests.

I hired a chef and we started developing recipes for plant-based dishes that could be used for the Jumpstart program. Back then, plant-based recipe options were far more limited, so we had to do much of our work from scratch. We even developed our own original recipe for a plant-based mayonnaise after searching the internet and finding nothing. We felt like pioneers, building a brand-new, groundbreaking program for people to reclaim their health. Once we had developed enough delicious recipes to support a meal plan, we decided to undertake our first Jumpstart.

We delivered this ten-day Jumpstart to about twenty people in the town of Wake Forest, just north of Raleigh, North Carolina, and we were shocked by what happened. Total and LDL cholesterol levels crashed by more than 20 percent on average, with many participants eliminating or substantially reducing their medication use.

This experience motivated us to undertake more Jumpstarts, this time with employees from a sporting goods employer in our small hometown west of Durham, North Carolina. We achieved the same remarkable results we had seen in Wake Forest. And it was around this time that my wife, Kim, jumped in with both feet.

Although she would never say this about herself, I can say from my years of observation that Kim is something of a culinary genius. She had originally intended to go to culinary school rather than attending a four-year university, but her father insisted she go to a four-year university. He didn't respect the idea of a culinary education, and as a former dairy farmer, he also didn't appreciate Kim's interest in plant-based nutrition. His views, however, in no way diminished her passion for food and healthy living.

Kim and I first met when we were only sixteen years old and married seven years later. It was in our early twenties that Kim started diving deep into the plant-based world, where she discovered a universe of culinary opportunity. She loved trying to recreate the traditional "comfort" foods we both loved in plant-based form. Since that time and up to the present day, researching and developing new culinary concepts has been her favorite activity, and over the years, she has published three highly regarded cookbooks. She also is an accomplished educator and loves sharing her knowledge with others.

Our initial chef had done a wonderful job, but she had just married and left the area to start her new family. Kim was happy to jump in to take over our culinary development. Motivated by our early discovery of the rapid healing power of a plant-based immersive experience, we began working on a broader vision I had developed for a community-based campaign strategy. This strategy would utilize a more scalable form of our immersion program, a new plant-based foods business to provide culinary support, and a network of support and advocacy groups led by local health advocates and other community leaders.

I decided to implement this strategy in nearby Durham and began reaching out to healthcare organizations that I thought could help to anchor a community campaign. I quickly learned, however, that this strategy was going to be harder to launch than I had anticipated. It required more capital than we had available, the logistics of a local rollout were complex, and I could not seem to find any sympathetic ears in the healthcare space. I became increasingly stressed and dejected, which is when my life took an unexpected turn.

After a surprisingly well-received presentation by my father and Dr. Esselstyn on the floor of the Kentucky House of Representatives, I secured an opportunity to work with the sponsor of that presentation, Representative Tom Riner, to help draft legislation that would have laid the groundwork for launching our community campaign model across Kentucky. It outlined a pilot strategy in the eastern part of the state to demonstrate the potential of a larger rollout, with the requirement that a successful pilot would lead to an opportunity for further discussions with government officials about a statewide campaign, focused initially on Kentucky's most underserved communities. Importantly, we did not ask for any money. We would find funding elsewhere and only wanted to demonstrate a powerful truth as a prelude to these larger discussions. Who could have been opposed to this? As it turned out, powerful industry interests did, and they killed our legislation behind the closed doors of the committee process.

Given our financial and other challenges, this could have been the end of our journey. But instead, it motivated my vision for a story-driven documentary film to make crystal clear to the public how our health message has been suppressed by powerful interests for decades. I envisioned a storyline that would start with our legislative failure in Kentucky and then run through our hometown in North Carolina, where we would do what I had wanted to do in Kentucky. With powerful data in hand, we would return to Kentucky for a dramatic showdown in its state legislature.

I reached out to people who knew how to produce films and could teach me the same, including John Corry, who had produced *Forks Over Knives.* John (who has become a close friend) taught me the ins and outs of filmmaking, then humbly stepped back into a producer role, giving me the space to write and direct my first film, *PlantPure Nation.*

Released in theaters in over a hundred communities in 2015 and subsequently distributed by Netflix, *PlantPure Nation* was well received. (It's now playing on Amazon Prime and Apple TV.) We soon got to work leveraging the movie's ending call to action to launch an international network of social support groups we called "Pods," to plant the seeds of change in local communities. It was at this time that I founded the nonprofit PlantPure Communities. Led and supported by a passionate and talented team, this organization increased the global reach of our network, which eventually involved hundreds of thousands of people.

The other leg of our strategy—the provision of healthy plant-based foods—became the focus of another organization I founded. Supported by a separate but equally passionate and talented team, our PlantPure foods business launched online with twenty frozen entree options. Publix Super Markets and a few other retail chains took notice, which enabled us to jump offline into the brick-and-mortar retail channel. We were becoming increasingly confident in our food strategy, which we believed could not only support the community model I had initially envisioned, but also create an economic engine to support these larger efforts.

Because the sales cycle in the retail sector is long, sometimes several years of repeated efforts to get products selected in a once-a-year category review, we had to work persistently to expand this business. By the early spring of 2020, after just winning a large commitment of new business, including from one of the largest supermarket chains in the U.S., we were finally poised for profitability later in the year. Our timing was awful, however; within just a few weeks of learning our good news, the COVID pandemic shutdowns hit, throwing our new retail customers in disarray. They canceled our newly won business, which forced our company into a financial ditch.

We somehow stayed afloat, but we found that the retail market for plant-based products was becoming increasingly competitive. Companies like Beyond Meat and Impossible Foods had developed highly processed plant-based products that were not only divorced from the whole food, plant-based message of my father, but also expensive. Consumers started losing interest in these processed food products, which accelerated in the inflationary days coming out of the pandemic. Unfortunately, category managers at retail chains had associated "plant-based" with these kinds of vegan products, so the whole plant-based category took a hit. And those companies that could continue playing in this space were far larger than us, with vastly more resources for in-store product promotions.

Seeing that we could no longer compete in the retail channel, we set to work on a new strategy. Leveraging our prior experiences and expertise, we decided to focus on an online, scalable version of our immersion program, supported by a new line of dry-pack meal starter products deliverable to consumer doorsteps.

Around this same time, the nonprofit secured funding for production of another movie, which we titled *From Food to Freedom*. This documentary film told the story of a live-in plant-based immersion program at a rented home in Greensboro, North Carolina, where we showed how type 2 diabetes could be reversed within ten days using a whole food, plant-based diet. We also explained the connection between nutrition and immunity. Our goal was to make the film freely available to maximize its public impact at a critical time on the heels of the pandemic, while capturing email addresses and new followers we could then leverage to expand our donor base.

Working with a marketing partner, we undertook a "sneak peek" release of this movie in the spring and were thrilled when over a hundred thousand people watched the film, many of whom provided strong positive feedback. This response motivated us to enter our film into festivals, where we won ten awards and honors from California to Europe, including "Best Documentary Feature Film" and "Best Inspirational Film.

We were hopeful as we prepared for our general release that summer. We began by promoting *From Food to Freedom* on social media, only to immediately discover that our film trailer had been restricted by the social media platform we were using. Working with our marketing partner, we lodged an appeal but were unable to get the restriction lifted.

We then uploaded our film to another platform, and our view count took off. Paranoid, I watched the number closely, waiting for the other shoe to drop, and sure enough, that's what happened. Our view count skidded to a near-halt in one day, a sign that this platform had decided to point its sharing algorithms away from our film.

Who knows exactly why our movie was suppressed. But I suspect it had something to do with the film's criticism of the pharma industry, which contributes significant ad revenue to social media platforms. When you watch *From Food to Freedom* (accessible on YouTube), you can see why a drug company would be displeased. Our film was compelling because we told real-world stories of healing, then we let our immersion participants express their feelings about the misleading information that had condemned them to a debilitating illness. This was far more powerful than commentary from "talking head experts," which is how information is often conveyed in documentaries.

Not surprisingly, the plans involving the new online immersion strategy and the movie, which showed how people can rapidly heal in an immersion process, suddenly collapsed. We could no longer support our staff at the nonprofit and subsequently handed off the Pod network to the Physicians Committee for Responsible Medicine in Washington, D.C. I lost my salary as well, and without a salary from either of our organizations to sustain us, Kim and I went into full panic mode. We hurriedly sold our house and moved to the western North Carolina mountains, where we could offer live-in immersion programs to cover our basic living costs and to buy more time to work on the vision that had motivated us from the beginning. We now live in a beautiful, small, idyllic mountain town, with access to a facility for our immersions and with many new friends in what I think must be the world's friendliest community.

Although stressful beyond anything we had expected, this shift in strategy at least gave us the additional time to keep our dream alive, and it seems closer than ever to becoming reality. We now find ourselves on the precipice of an opportunity to demonstrate the community model I envisioned for Durham so many years ago, and in an even more impactful way. This opportunity is anchored by a local network of publicly funded community health clinics, and we have recruited additional partners with the capacity to help scale this model nationwide once it's validated. As I write this, however, we're working on securing the capital we require in PlantPure, which we've repositioned to pursue this opportunity, but I'm not sure we'll succeed. It's hard raising money today for a plant-based project like ours. Many individual investors and investor groups have lost interest in the plant-based idea.

Sadly, I think our movement has flatlined and become somewhat irrelevant, at least in the U.S. I know some advocates in our community will take offense to that statement, but I would encourage them to take a step back to see the big picture. I can see this picture, having had a front row seat to my father's research and then engaging in strategies with others over a span of more than fifteen years to promote the health truth he discovered. I recounted the journey we've taken to gain some credibility with you because in the remainder of this book I'm going to say some things that will resonate with some people in our community, but will feel provocative to others.

My commentary begins with this observation that our movement has stalled. Around the time of our *PlantPure Nation* movie release, interest in plant-based nutrition was exploding. This interest drew the attention of investors and entrepreneurs who picked up the "vegan" flag and hijacked the plant-based movement. As I recounted earlier, I know firsthand, through our PlantPure foods business, how junk food vegan companies undermined public interest in plant-based nutrition. These companies ditched the health message of whole food, plant-based nutrition so they could sell more of their engineered fake animal products, and by virtue of their dominant role in the marketplace, they redefined plant-based as expensive, unhealthy, and heavily processed. This has not been an effective sales message for the movement they hijacked.

The people who are selling the real versions of these vegan products—actual meat, cheese, milk, and eggs—have clearly offered a more successful message. The biggest food trends today are the paleo and carnivore diets, which are really facets of the same animal-based diet.

To make matters worse, consumers remain lost in a sea of conflicting health claims, our healthcare system and mainstream media continue ignoring the plant-based nutrition message, and there is little hope that public policy on this matter will change anytime soon given recent pronouncements from our policy leaders about what constitutes optimal nutrition. Of course, some of what we're seeing today is simply a product of ignorance, but there is more to the story.

I remember when my family first learned that an industry group had been formed to monitor and attack the credibility of my father. I remember the clicks and other sounds, including muffled voices, we heard whenever we called my parents. I remember one call from a third party that came into my office in North Carolina, simultaneously answered by my mother at their home in upstate New York. This was several years prior to his publication of *The China Study*, and I remember wanting my father to finish his book as quickly as possible so he could get his research and ideas into the public sphere; there is nothing more dangerous to a person than being the chief advocate of a hidden truth disruptive to the rich and powerful.

What my father experienced, however, pales in comparison to what is coming, and we'll need to understand this future in order to craft a strategy that enables our message to win out. But before discussing strategy, we need to start with a more foundational question: what is the message we want to communicate? Every movement must begin with its rallying cry.

CHAPTER 15

WHAT ARE WE SAYING?

Before diving into this issue of messaging, I want to acknowledge the efforts of all the people working to advance the health, environmental, and ethical arguments for a plant-based world. I salute the people who have been fighting this good fight, many of whom have become my good friends. My comments here are not intended to show any disrespect to them or to undermine enthusiasm for our cause. Quite to the contrary—the ideas I'd like to share are meant only to stimulate some new thinking so we can formulate more effective approaches to engage in together.

We need more effective approaches. As I explained earlier, our movement has stalled. A Gallup poll conducted in 2023 found that only 1 percent of Americans identified as vegan and 4 percent as vegetarian, a decrease from previous years.[55] But even more informative may be a research study from 2014 that estimated 10 percent of U.S. adults were *former* vegetarians or vegans.[56] Based on my own experiences, I believe this 10:1 ratio of former vegans/vegetarians to current vegans/vegetarians still holds true today. If anything, this ratio is now even greater. Most people who experiment with plant-based eating end up reversing course. This is a very important point, which lays a foundation for what I would like to say about our messaging.

I understand the people we are trying to reach because I also grew up on an animal-based diet. We had side veggies with every meal but usually had animal-based products as the centerpiece of our meals. My family also kept horses when I was growing up. Spending time with these beautiful, sensitive animals, and thinking that meat was healthy, I thought someday I would move west to become a cattle-ranching cowboy, where I could ride through pristine open spaces during the day and curl up at night next to a fire, under a dark canvas painted with starlight. This dream dissolved in my late teens as my father made his science discoveries, so in place of the cowboy vision, I started thinking about produce farming.

I say all this to underscore that I empathize with people who are consuming an animal-based diet. Like them, I loved the flavors and textures of the food I was eating, which is why I was so surprised by the new dishes Kim created early in our transition to a whole food, plant-based diet. She worked hard to create mouth-watering recipes that even the most ardent meat-eater can enjoy. (And as the dishwasher in our house, I found cleaning much easier as well!) After experiencing both ways of eating, I would never go back to eating animal products, even if they were found to be as healthy as plants, because a plant-based diet is just more delicious. I also have come to love animals, and for that reason as well, I would never go back.

A plant-based diet can, indeed, be joyful, and that's a point we should emphasize to others. Herein lies the first part of our messaging that we must change if we want to build a plant-based world.

DIETARY EXTREMISM

My father is the one who first coined the phrase, "whole food, plant-based." He recognized that processed plant-based foods could be as unhealthy as animal-based foods, so he wanted to emphasize the value of whole foods in their natural state. Ironically, many people in the plant-based community have taken this concept to an extreme and turned it back into a reductionist idea. It seems we can't help ourselves; we always want to stare at the tree in front of us rather than standing back to behold the beauty of the forest.

An example of this reductionism is a notion that has gained strong traction in our community: that a whole food, plant-based diet must be devoid of any salt, sugar, and oil. This way of eating is touted as an "SOS-free" diet (salt-, oil-, and sugar-free). Both my father and I contributed to this line of thinking in the early days, when we sometimes spoke about the most natural form of the plant-based diet—what we sometimes defined as optimal—as one consisting of whole foods free of refined ingredients, including oil, salt, and sugar. My dad had focused his research on animal protein—not oil, salt, and sugar—but it seemed sensible to talk about a natural whole food diet this way. With hindsight, I wish we had been more careful in our wording.

I don't have an issue with the absence of refined oil, aside from minor uses such as lightly seasoning a pan, because oil is not essential to creating delicious plant-based dishes. Some savory and sweet notes, however, are critical to creating appealing flavors. When I speak to audiences about this SOS-free concept, I often encourage people to imagine a plate of brown rice and an assortment of colorful cooked veggies on top. Most people would expect this dish to include some flavorful sauce on top or infused throughout, but if you are on an SOS-free diet, that's a "no-no" if the sauce has even a crystal of salt or sugar. This is an extreme restriction considering what's actually on the plate. According to my father, there is no research anywhere that suggests a modest amount of flavoring has any detrimental effects whatsoever in the context of a whole food, plant-based meal. I once asked him to review the research on salt to see if he could find any studies showing that a reduction in daily salt intake below 1500 mg a day to the sodium levels naturally found in plants has any discernible impact on our health, and he could find nothing. This was not surprising to him. Blood pressure is mostly associated with levels of salt intake higher than the 1500 mg limit recommended as an optimal level by the American Heart Association. Also, my father has found that plants intrinsically help to regulate blood pressure, and that blood pressure can be impacted by other important lifestyle factors in addition to excessive sodium, such as physical activity, stress, body weight, and age.

Of course, many of us consume too much salt, and we should do our best to minimize our intake. Fortunately, it's possible to create many delicious dishes on a low-sodium diet. Our taste for salt is malleable, which means we can reduce our salt intake over time without losing the savory notes in our dishes that help to create their flavor. But while it's a good idea to limit our salt intake, we need not eliminate salt completely. According to my father, this would be not only scientifically unfounded, but also reductionist. Rather than seeing the whole dish and the tsunami of nutrients in it, we fixate on a meaningless, tiny fraction of the dish. We obsess about it to the point of forcing ourselves to eat less flavorful food, which makes living a plant-based diet difficult, and for some, impossible.

This same argument also applies to small amounts of sugar. For most of us, a small amount of sugar is of no consequence to our health. That being said, it's important to acknowledge here that sugar (and salt) can be triggers for people who struggle with overeating. It's also important, though, to consider what is being triggered. A little flavor that makes eating an oil-free veggie stir-fry joyful is not the same as an excessive combination of salt, sugar, and oil triggering someone to binge on an unhealthy processed food.

Fortunately, processed junk foods can be removed from the home, which can be an effective solution—perhaps more so than the radical step of removing flavor from healthy foods. We've found that people struggling with overeating who think they must consume flavorless plant-based foods often fall off the wagon, even if for short bingeing periods. And of course, there are often other issues underpinning the drive to overeat, which may require resolution in other ways.

Another form of dietary extremism can be found in the common admonition to severely limit or even eliminate from one's diet plant-based foods with whole fats, such as nuts, seeds, avocados, soy, and coconuts. We should of course stay away from oils in a bottle because they are highly processed and unhealthy (and not necessary to producing delicious plant-based meals). However, these oils are not the same as the fats present in whole plant foods. As my father explained in an article reprinted in Kim's cookbook, *The PlantPure Kitchen*, nuts and seeds have important nutrients, including fat, and when eaten in moderation, as Nature presents them, they promote heart health and may even cut the risk of cancer.

The paranoia about whole fats in plant-based foods is reflected in a phrase famous in the plant-based community—"The fat you eat is the fat you wear." This slogan is just a reincarnation of the same logic used by the dairy industry, who told us to drink milk so we could "wear its calcium" in the form of stronger bones. As my father explained earlier in the book, biology does not work in such a simplistic way. As it turns out, our bones lose calcium when we consume foods high in animal protein, which makes them more prone to fracture in our later years. Similarly, the fat in plant foods won't negatively affect your health when eaten in moderation as part of a varied, whole food, plant-based diet. Drawing a line from eating whole plant-based foods with fat to conditions like obesity misses the seemingly infinite biological complexity in between and is just another example of reductionist thought.

Related to this general concern about fat is the myth of saturated fat. As my father recounted earlier, saturated fat became the "bad guy" years ago as an industry strategy to divert attention from animal protein. My father has often pointed out in his talks and writings that the chemical structure of saturated fat makes it mostly inert in the human body. This is not well understood by people with insufficient training in biochemistry, so in addition to minimizing or eliminating nuts, seeds, and avocados, many in the plant-based community also stay away from coconut, which contains saturated fat. In making coconuts villainous, we are parroting a claim made years ago by people wanting to protect the sacred reputation of animal protein.

These arguments—that higher-fat plant foods make you fat and that saturated fat from whole plants is harmful—are also flawed for an even more fundamental reason. For millennia, we lived immersed in Nature, and in those natural conditions we developed a physiology that is optimized for eating plants—and very different from that of carnivorous animals. Our optimal diet comes from Nature, and anyone who has spent time outdoors knows the natural environment is full of foods like nuts, seeds, soybeans, avocados, and coconuts, especially in the lower latitude areas where early humans thrived. These plant-based foods must have been appealing to them; early humans would have loved their flavor and benefited from the satiety and energy they provided. I doubt they carried fat meters around to identify and then avoid such delicious and satiating foods. Simply put, these foods are natural to our biology. We often forget that the healing power of a plant-based diet comes from Nature, not from experts who make scientifically unfounded and reductionist pronouncements about which plant-based foods are good and which ones are bad.

All of this—SOS-free restrictions and the elimination of foods with whole fats—makes it far more difficult to create plant-based dishes that can give joy to mainstream folks. Over the years, I've met more people than I can remember who were taught by someone else an extreme version of the plant-based diet, and who either immediately spurned the diet or tried it and couldn't continue. Unfortunately, the idea that a plant-based diet is sacrificial rather than joyful has become a dominant perspective in our community. Even the term "SOS" suggests something extreme. People who eat chicken nuggets and cheeseburgers are not going to look at our community and willingly jump into a diet called the "SOS-free" diet (talk about poor branding!).

How are we going to build a plant-based world with such unpopular (and scientifically invalid) ideas, which are sometimes directed judgmentally against people not following the so-called "perfect" form of our diet? This is an exclusionary approach, in my opinion, because it pushes people away from a path of healing. I realize proponents of this kind of dietary extremism may not see what they are saying in this light. But regardless of intention, the effects are undeniably counterproductive because they impede our mission of building a plant-based world.

DIETARY CONFUSION

Selling an idea is all about crafting and communicating your message in a way that resonates with your audience. Content is important, as is the manner in which content is communicated. The most effective communication is simple and consistent, devoid of conflict or complexity that could confuse the minds of an audience.

Unfortunately, we have often failed to communicate the empowering health benefits of a plant-based diet in a simple and consistent manner. As a result, our community is mired in confusion. Kim and I see this firsthand when we host people at our live-in immersion events. They arrive confused, feeling exasperated by all the tedious and conflicting information they have heard. Kim and I try to untie these knots so we can support people in sustaining a plant-based lifestyle.

Although often well-intentioned, "expert" voices in our community sometimes confuse people because they see the plant-based diet through a reductionist lens, fixating on its parts so they can prescribe what they see as the ideal form of the diet. For example, we are advised to eat more of certain whole plant-based foods—"superfoods"—because of their supposed nutrient profile, and therefore, to eat less of other foods of supposedly lower nutritional quality. It's not enough to eat a variety of plant-based foods; now we must go through the tedious process of figuring out which of these foods are better for us than others and somehow adjust our consumption accordingly. We also hear advice to eat certain whole plant-based foods together and conversely to eat other foods separately, to take supplements, to consume an all or mostly raw diet, or if we cook, to cook some foods and not others (like tofu and breads)—the list goes on and on.

Of all these tedious recommendations, the worst may be the advice to take supplements. There may sometimes be situations where supplements can provide some benefit, but they are usually unnecessary and can even cause complications. My father discussed these points earlier in the book, so I will only add here that supplements are just plain confusing. Once we open the Pandora's box of supplements, there's no end to the snake-oil concoctions people can sell to us. And how can we possibly judge the veracity of their claims? I think my father's suggestion to focus on eating a variety of whole plant foods rather than on swallowing pills is probably all we need to remember.

Whenever you hear a self-proclaimed expert making the kinds of claims I just shared, you can most often discount them. This raises another point, however: it can be hard to discount such messages when they come from people with medical degrees.

I am a fan of physicians because most of them chose their career so they could heal people. Some physicians might have gone into the profession primarily for money, but I think most became doctors because they like the idea of making sick people well—such a noble impulse. I also believe physicians must be at the front lines of our strategy to build a plant-based world. They have a valuable bond of trust with their patients, and because of that bond, they can drive interest in the plant-based message better than anyone. I respect physicians and understand the leadership role they can play in our movement.

Having said this, I want to caution against submitting to advice merely because it comes from a person who earned an MD. Physicians generally do not receive substantive training in nutrition in medical school, although they certainly have other expertise that is deserving of respect. Within their practice areas, they have expert knowledge of disease symptoms and of the pharmaceutical and other medical interventions and procedures used to manage these symptoms. They also understand the "standard of care" used in our medical system. But while important, this expertise is an entirely different kind of knowledge than the knowledge gained from the kind of deep and wholistic biological research performed by my father.

Unfortunately, some people in the medical community have used their degree to claim expertise they do not have, perhaps to sell a program, to gain social media followers, to motivate donations to an organization, or simply to gain a reputation as an expert. This has given rise to pronouncements about health that are conflicting, confusing, and scientifically unfounded.

Kim and I recently held an immersion event at our location in North Carolina and invited my parents to attend, so they could see firsthand what we do in an immersion. Over the course of two seminars, I told my father's life story. Afterward, I invited him to come up front for a question-and-answer session with the audience. A smart and humble physician, who retired from his oncology practice after learning about plant-based nutrition, asked a penetrating question: why does animal protein promote cancer while plant protein does not? Sitting in front of him, a little slouched in his chair, was the person who has done more research on this topic than any human alive, and most likely any human who will come after him. My father sat there silently, his head bowed. He seemed lost in thought, then after a moment, he lifted his head and said (I'm paraphrasing), "I wish I could give you a full answer, but the truth is *I don't know*; biology is so complex. Notwithstanding our ignorance of the reason, however, we know from observation that what I've said about animal and plant protein is true." This humility is what enabled my father to see through his biases to discover a truth destined to change the world.

I say "destined to change the world" because that's how truth operates, especially one as important as the truth of plant-based nutrition. A truth like this can't be suppressed forever. But while this change is written into our future, how quickly it comes to fruition will depend in part on how we communicate our message to mainstream society.

The message Kim and I communicate to the people we work with is this:

We heal ourselves when we eat a plant-based diet, eating from all parts of the plant, in a rainbow of colors, and infused with flavors that make eating this way joyful. We need not use refined oils in our recipes, but there's no problem with small amounts of salt and sweeteners to create flavor; we should not obsess about details that are inconsequential in the context of the whole. We can enjoy our food and tune out the conflicting and tedious messages we hear online and elsewhere, relaxing in the knowledge that we are already doing what matters most for our long-term health.

If we can communicate our health message this way—simply and with joy—we have taken the first step required for launching a movement to build a plant-based world. I think we can then take the next step, which must be to broaden our message even further, especially in a way that enables currently disparate groups to work together in common cause and that encourages the greater participation of young people.

CHAPTER 16

BROADENING OUR APPEAL

Many people adopt a plant-based diet out of concern for their personal health, and rightly so. This concern should remain the foundational element of our argument, especially in light of a new threat from the pharmaceutical industry.

Just a few days prior to the time I'm writing this, U.S. President Donald Trump announced that the government had negotiated steep price discounts on several major weight loss medications. These drugs enable people to lose weight by suppressing their appetites—so they eat the same unhealthy foods, just in lesser amounts. For people who are obese, and who may need a kickstart toward adopting a healthier lifestyle, maybe these drugs could provide some benefit—but that's not how they'll be used. These drugs may become commonplace soon, at least in America, and many of the people taking them, if they aren't suffering through intolerable side effects, will likely take them on a long-term basis in lieu of making healthy lifestyle changes.

To overcome this threat from the drug industry, we need to

continue promoting in our movement the health benefits of a whole food, plant-based diet, using the science-based arguments my father and I have shared in this book. But while the health argument should remain at the core of our messaging, we must recognize that many people are inclined toward a plant-based diet for different reasons: a deep love for animals and/or a concern about the larger environmental impacts of our food choices. Many of these are young people.

Whenever Kim and I host an immersion event, our participants overwhelmingly come from the forty-five-plus age group, with many baby boomers in attendance, which makes sense given our focus on health. We love working with these people and count many friends among them. But we always have this nagging feeling that we're not engaging with the demographic most important to our future: young people.

Kim and I just welcomed our first grandchild, a beautiful boy. When we look into his eyes, we see innocence, manifested in his ability to see the world as it is, without bias. He spends all day exploring, in a constant state of wonder. Kim and I relish our time with him and think often about what his life will be like in the decades ahead. This motivates us to renew our efforts to live our lives with purpose, doing whatever we can to bring more healing into our world.

The innocence we see in the eyes of our grandson is a gift we all share early in life, but one we start to lose with the passage of time and the accumulation of experience. Fortunately, however, we retain some of this perspective into early adulthood, as the innocence of our childhood morphs into a more informed idealism.

This more idealistic perspective is what I love about young people, who as a group often express great concern about social inequities, the lives of animals, and the state of our environment. They may not have the health concerns of older generations, but they often care deeply about these other issues.

Young people also have energy. As an older person, I can attest to the fact that life can wear you down. In our early years, we begin with a deep reservoir of energy, which combined with a more idealistic perspective, can motivate us to pursue bigger ideas for remaking our society. This is why young people are critical to the success of any mass movement, and why we must expand our plant-based message beyond health.

A MORE YOUTHFUL MESSAGE

I believe we need a message for our movement that is more inclusive and expansive. Our message also needs to be rooted in the idea that our goal is not only a plant-based world, but also one that is kinder and more connected. In the New Testament, we hear the charge of Jesus to love "the least of these." In saying this, he was charging us to love those whose suffering is greatest due to life circumstances. I think we have done a poor job expressing this kind of love in our plant-based community.

Years ago, a marketing consultant once told me, "If you want to succeed in selling a wellness product or service, you need to target the 'Whole Foods Market demographic.'" By that, he meant the type of people who shop at Whole Foods grocery stores—

people who are more affluent, more educated, and supposedly more health-conscious. When he said this, I felt disgust, because health is not something we ought to promote only to certain advantaged groups in our society.

Over the years, we have led or supported successful immersion programs involving diverse groups of people, many of which don't fit this suggested demographic. These experiences have reinforced my belief that we should ignore the advice of marketing consultants who tell us to target our outreach to more advantaged communities. We need to reach out to *everyone*, including those segments of our population who have the same human potential, but who are the most under-resourced and most plagued by chronic disease. I believe that, no matter the challenges, we must bring healing to all our brothers and sisters, whoever they are and wherever they live.

And while we're talking about growing our circle of compassion, I think we should extend this circle to include animals . . . all animals. Though we love our pets, we often turn a blind eye to the suffering of the animals whose flesh and lactation fluids end up on our plates. Most of us today are so isolated from Nature that the only animals we ever connect with are the dogs and cats who live with us. Yet, cows are smart, gentle, and, yes, loving creatures. That's apparent when a calf is ripped from its mother following birth; both the mother and calf will often wail for hours on end. And what about pigs? Have you ever gone down an interstate highway and passed a large truck jammed full of pigs, with their snouts and tails visible through the little holes in the walls of the trailer? Did you ever consider the fear they might be feeling on their way to the slaughterhouse? Or the pain felt by chickens in suffocatingly crowded pens, who have been bred to be so heavy they can't stand on their own two feet?

Kim and I have a wonderful companion named Molly, gifted to us by our two daughters. She is an incredibly affectionate dog, who seems to have come into existence for the sole purpose of loving the people around her. Kim and I spend much of our time with Molly, and we see in her a profound inner beauty and a loving spirit that comes from the same source we all come from. But why should we not see the same in all animals?

As my father's research has shown beyond any doubt, we harm ourselves when we eat animals. And we do this while also causing suffering for the animals we consume. Most people still don't understand the health benefits of a plant-based diet, and they eat the food they were raised to eat—food I also ate earlier in life—so I'm making no judgments here. What I'm advocating is that we raise awareness of the suffering caused to both people and animals when we eat animal products, but always with humility, never with judgment.

Finally, if our goal is a kinder, more connected world, we can't ignore the way we view and treat Nature more generally. We often think we are above the natural world. We live apart from Nature, in our homes, workplaces, and cars, and in this separateness, we forget that we are only a small part of the huge, interconnected web of Nature. Perhaps we can take a lesson here from the many Indigenous societies around the world, who have always been much more advanced on this point.

We have harmed Nature in so many ways, and much of this happens through our food choices. There are uncountable examples of this, more than we know, but I will focus on just one. The Food and

Food and Agricultural Organization (FAO) of the United Nations issued a report in 2013 claiming that animal agriculture is responsible for 14.5 percent of all anthropogenic greenhouse-gas emissions.[57] A more recent analysis led by Dr. Atul Jain estimated that animal-based agriculture accounts for roughly one-fifth of total human-caused greenhouse-gas emissions worldwide—significantly higher than the earlier UN estimate and about twice the emissions associated with plant-based food production.[58]

And the real impact of animal agriculture on our climate is likely far higher. As the latter study noted, most assessments do not include the opportunity costs of lost carbon-sequestration capacity on agricultural land, most of which is used for raising animals, and its authors called for further research to quantify these impacts.

More specifically, humans have already cleared nearly half of Earth's original temperate and tropical forests, largely to create grazing pastures or to grow crops for animals. These forests once stored enormous quantities of carbon in their biomass and soils. In their place is an animal agriculture system that not only emits carbon dioxide but also produces large amounts of methane and nitrous oxide—greenhouse gases far more potent than carbon dioxide. A global shift toward plant-based diets would sharply reduce these emissions while freeing vast areas of land for restoration. Research suggests that restoring roughly 0.9 billion hectares of forest—an increase in global forest cover of about 25 percent—could capture on the order of 205 gigatons of carbon, which is equivalent to about 750 gigatons of CO_2. This would be enough to reduce atmospheric carbon dioxide by roughly one-quarter, or about two-thirds of the excess CO_2 humans have added since the Industrial Revolution.[59]

In short, dietary change is a critical component of resolving the climate crisis.

As noted, this is but one example of our many insults to Nature. By changing our food choices, we also could slow and then stop biodiversity loss, save precious topsoil, improve water supply and quality, heal our marine environments, and more. Many books have been written on these topics, so I'll leave you to these other resources if you want to go deeper; the simple point I'm wanting to make here is that the food we choose to put on our plates has a powerful impact on the natural world.

Everything I've suggested could be building blocks of a larger, more compassionate message that would enable us to unify disparate groups focused on health, social inequities, animals, and the environment into a unified movement for change. I also believe we could appeal to people who aren't plant-based but yearning for a different kind of world. We live in a fractured, chaotic time, when greed, disrespect, and even hate often seem the norm, at least according to the mainstream media. I believe most people want the opposite of this, and as I said at the outset, I believe this more transcendent messaging would appeal especially to young people, whose participation in our movement we desperately need.

CHAPTER 17

MESSAGE TO ACTION

Compelling messaging is a starting point for any movement, but equally important is strategy. Before sharing my views on this point, I need to preface what I'll say by again recognizing the people in our community providing information, products, and services that enable people to learn about, transition to, and sustain a plant-based lifestyle. I'm a fan of these passionate people, who are working hard to make unique contributions to the world.

We have lots of these folks in our community, many of whom we think of as leaders. And indeed, these influential voices have helped to motivate greater public interest in plant-based nutrition. Their strategies, though, are often individually pursued and confined to the digital world. While helpful, I don't believe this kind of approach, alone, will lead to the kind of transformational change we're seeking.

I'll delve into this argument further, but only after taking a short detour to examine what we're fighting. I mentioned earlier that the darkness my father battled in his career is but a taste of the future that's coming. We need to understand what we're up against if we are to devise an effective strategy for our movement.

THE TRUMAN SHOW

In the past, if an industry group wanted to oppose my father, they would often need to meet in person to develop and coordinate their strategies, and sometimes even in airports, as my father recounted in past writings. Then they would need to use landlines, fax and copy machines, hit pieces published in traditional media channels, campaign contributions to politicians, and so forth to exert their influence. No longer. While those tools were, and still are, powerful, they pale in comparison to the power that can be wielded through today's new technologies.

I found *The Truman Show* with Jim Carrey an unsettling film. I think I felt this way when I watched it because the world Carrey's character inhabited in that film, which was completely manufactured, bears some resemblance to the technology-driven world we inhabit today. The town square, and any other place where, in times past, people would gather to talk and share information, has been largely replaced by digital platforms accessed through our smart phones and computers. We peer into glowing screens to talk and learn new information, which creates a ripe opportunity for powerful forces to replace reality with fiction, controlling what we see and think. There are many ways this can happen, but for purposes of illustration, I'll share just two examples.

Conversation happens through social media in a pretty

straightforward way. People post some sort of content, then discussion often follows in comment threads. This discussion can now be manipulated by a new technology so powerful that it almost feels magical: artificial intelligence (AI). People with money and an agenda can use AI to influence online discussion through armies of fake accounts called bots. These AI agents seem like real people and can converse in real time to amplify or sabotage a message. So, for example, when a plant-based nutrition advocate uploads a video or posts information, a network of bots can flood the comment section with criticisms and false information, while a parallel swarm showers carnivore diet posts with praise and spreads more false information. To a casual reader, the result looks like authentic consensus, and to the platform's ranking algorithm, it looks like popularity, triggering still wider distribution.

This tactic is extraordinarily cheap and almost impossible for ordinary users to trace. Because each bot appears to be an independent individual, the real sponsor remains hidden. And even when platforms detect and remove these bots, new swarms can be spun up overnight for pennies.

I should note here that I think AI can be used for immense good, such as for creating new technologies to address the problem of climate change. The problems that come with AI in my view are mostly caused by people driving its development in a careless way and by people using its power for the wrong reasons.

In the future, AI will drive much of what happens online, but I would be remiss if I didn't also mention the role of social media influencers. Many of these people are authentic in their beliefs but are nevertheless often influenced by the same forces that employ

AI in nefarious ways. Industries trying to protect their financial interests have become adept at targeting and then paying influencers who communicate ideas aligned with those interests.

Within a digital world of AI-generated fake people and of real people influenced by corporate interests, it's becoming increasingly difficult to know what's true and what's false. By using technology and capital to create a world divorced from truth, the dark forces who profit from our suffering gain more power by the day—and we'll never win if we play their game. The good news is that there is another game we can play, but we need to think outside the box to see it.

A GAME WE CAN WIN

I'll just cut to the chase here because I think I've already laid the groundwork for this statement:

> *To build a plant-based world, we need to improve our messaging, then expand our reach beyond the digital world by engaging in our real-world communities. We need local leaders who can develop and spearhead strategies that empower people to connect with others in the old-fashioned way, face to face, in family rooms, community meeting rooms, worksites, places of worship, and physician offices. We also need to provide real-world resources, especially healthy food options and social support, to people who have committed to a plant-based lifestyle.*

There is a time and place for online advocacy, but we'll never overcome the forces that resist us if we spend all of our time behind computer and smartphone screens. The digital world is far too easy to manipulate today, making it difficult if not impossible for us to discern what's real and fake, true and untrue. The industries and politicians who profit at our expense, however, cannot intervene in the relationships we have in our families, worksites, neighborhoods, and wherever else we come together. This is the real world we inhabit, and in this world, we are in control.

Our truth is one that can be easily actualized and spread among people who are connected. We're not talking about a difficult-to-prove truth in the realm of physics or mathematics, but about a truth that becomes immediately apparent whenever a person suffering from a chronic condition finds healing through a plant-based diet. If we can help people find such healing in our communities, then empower them to share this truth with others around them, all while providing the resources required for a plant-based lifestyle, there is no amount of money or political power that can prevent that truth from going viral.

I think there's tremendous promise in the local, real-world strategy I'm suggesting here. But unfortunately, I've seen too little consideration of this approach in our plant-based community, perhaps because this approach is far more challenging than a digital one. I know these challenges from direct experience over more than fifteen years pursuing such a strategy, often struggling in ways my team and I had not anticipated. But we've persisted, and as I explained earlier, we now sit on the precipice of an opportunity to fully implement the vision we started with—a wholistic community model, anchored by a local network of publicly funded community health clinics and supported by partners who have the capacity to scale this model nationwide. We unfortunately don't yet have the resources to take advantage of this extraordinary opportunity, so maybe it will be just a memory by the time this book is published, but if so, I'm hoping that what I've shared here will motivate someone else with leadership skills and access to capital to pick up where we left off.

CHAPTER 18

PLANTS AND POLITICS

I've always had strong political motivations, even in my youth. I first volunteered in my local congressman's campaign during middle school, licking stamps and sending out letters, collecting and organizing voter and campaign data, and doing the other mindless tasks often done by volunteers. I did the same in high school, and then in college landed an opportunity to intern in that same congressman's Washington, D.C., office. I took a semester off from my studies at Cornell University to do this, which then led to an opportunity to serve as his deputy campaign manager that summer.

In that job, I suddenly had access to information I had not known before, such as the source of our congressman's funding support. This explained to me discrepancies I had seen between his privately stated views and certain of his public positions. This information, combined with the superficialities of the campaign process, shattered my dream of making change through politics, and I quit my newly won position to return to school. This was the only job I have ever abruptly quit, but I felt like I had no choice. I couldn't do something that was not in my heart to do; I'm not very good at living a lie.

Notwithstanding this experience early in my life, I've remained interested in political issues, and I've developed a perspective that I believe could help to add a political dimension to our plant-based movement. People today are frustrated living under a system controlled by the rich and powerful, and I think there is an opportunity for our movement to shed some light on a more unified and empowering approach to governance.

A ROOTED POLITICS

I believe our political system is upside down. Most of the tax dollars taken from our paychecks are spent somewhere other than in our own communities. These dollars flow from us to Washington, D.C., and to a lesser degree to our state capitals, who then spend the money as they see fit. Some of this money is put to good use, but much of it is used to fuel policies that benefit powerful special interests.

When you think about it, our top-down approach to governing is similar to our approach to healthcare. In matters of health, we hand over our personal health to techno-medical "wizards," who, like the Wizard of Oz, project authority and assure us they know how to care for us. Yet, we now understand that the real power to change our health lies within us, and it's a power we control simply through the food we choose to eat. When we eat the right foods, our trillions of cells and all the other elements of our biological system work together in a way we'll never fully understand to naturally produce health. There is no conductor at the top; it's an entirely organic process where every part of the system, no matter how small, contributes to the health of the whole.

Just as we rely on techno-wizards in healthcare, we rely on our political "leaders" above us, who tell us they know how to fix what ails us. They promise solutions to win our votes but rarely deliver. And just as we can use food for true healing, I think we can engineer a new political approach that empowers people everywhere to heal our world in an organic, bottom-up way.

And healing is possible. There is not any problem we have in our modern society that has not been solved by some social entrepreneur somewhere, who has responded to local conditions using local relationships and resources to do something heroic. I challenge anyone reading this book to research a social ill of your choosing to see if you can find a solution that some social entrepreneur somewhere has demonstrated. I've never failed to find a story that inspires me.

Unfortunately, these stories are almost always the same. Their heroes often have difficulty finding the financial resources they require to sustain and expand their efforts, and almost never have the ability to share their successes beyond their own communities. These inspiring souls are like fireflies in the night—beautiful when lit up, but never casting enough light to illuminate the world. It's not their fault, but rather the fault of our upside-down system sucking resources out of our communities and sending them to distant places, where they can often be misdirected for the benefit of the rich and powerful.

I believe we need to develop ways to reroute a larger share

of the people's money through local communities, bypassing the corrupt people at the top. Imagine if our plant-based movement could help slash the cost of healthcare—the largest item in our federal budget. Along with other shifted spending priorities, we could redirect tax dollars to the places where we live to empower solutions to our most pressing problems. There is a lot to this idea that would need to be figured out, but I think that if we were able to land people on the moon in 1969, we ought to be able to figure out a transparent and accountable way to disburse tax dollars to social entrepreneurs and organizations working to heal their communities.

We also could empower these healing agents with information, especially information on best practices, to foster an organic cycle of mass innovation. If a group in one community achieved success in dealing with a difficult problem, they could upload that information to some sort of database accessible to everyone, with filters and ranking systems making it easy for people to find the ideas they're looking for.

Financial and informational resources flowing in the way I've suggested would be a kind of "social nutrition," empowering an organic healing process from the bottom up, driven by people helping their neighbors. Rather than living at the mercy of governmental overlords, concocting and forcing costly and mostly ineffectual solutions, we could live in a society where millions of people are empowered to bring about true healing.

I believe our plant-based movement could help make the case for this political argument. Showing how we can solve one of our most intractable problems—our epidemic of chronic disease—and doing so through a community-based approach would demonstrate the power of a political system turned right side up—*a true democratic republic.*

Perhaps then we could move past the blinding partisanship that has paralyzed our politics. I say "blinding" because there are splinters of truth everywhere—not only on whatever side we sit. Yet, when we emotionally cling to the talking points of our side, we can't see these splinters of truth elsewhere, which is necessary if we are to fashion larger, more transformative truths for building a better world. I believe the political idea I've shared here is an example of such a truth. And it's when we reach for these kinds of transcendent truths that we are able to bring people together. Indeed, when I talk to people on both the Left and the Right about the idea of shifting the locus of power to our local communities, I get an equally enthusiastic reaction from both sides.

Who knows? Perhaps these ideas could be used to add a political dimension to our movement. Maybe we shouldn't shy away from mixing plants with politics; people tend to get excited about movements that are political because they see them as opportunities for making the greatest societal change.

I'm sharing these thoughts, if only in a superficial way, to spark thought and discussion about how our movement could further transcend the simple question of what we should and should not eat. My father's stories earlier in the book were stories not merely about nutrition, but about a deeply flawed, top-down political and economic system that mirrors our current healthcare system. Maybe we could address both systems simultaneously through our plant-based movement.

CHAPTER 19

FINAL THOUGHTS

The vision I've shared is now lacking only one more element—perhaps the most important one. I've talked about macro-level ideas involving messaging, community-based strategies, and even a new political paradigm. But confining ourselves to a debate on a macro-level is not enough.

There is a unique perspective that underlies all I've discussed, which I call "heart-centered wholism." This perspective is wholistic because it sees the complexity of our world, and it's heart-centered because it seeks to cultivate compassion and connection. Because I'm talking fundamentally about a new way of seeing and engaging with our world, this is a change that must start within each of us. My hope is that we can each take a moment to reflect on our lives and to consider whether we are seeing the world through the lens of wholism and living with compassionate purpose.

Kim and I find motivation to live this way through our spiritual beliefs. We don't believe the universe is a random conglomeration of matter. Given the evidence to the contrary, that belief would require more faith than we could ever muster. Instead, we believe there is a creative and compassionate force driving the movement of our universe, which makes that movement intentional and profoundly beautiful. Even though we can't directly see this immaterial aspect to our universe and don't know where we're collectively heading, we can discern that our lives have meaning and that we have been gifted life so that we can evolve our souls, which we do through love. We can know this like we know the wind exists: We cannot see the wind, but we experience its effects.

I know many people reading this book may be atheistic or agnostic, and I'm not trying to persuade you to any particular faith tradition. I would like to invite you, though, to step outside the immediate material circumstances of your lives to see the world around you in its fullness—its complexity, connection, beauty, and inherent purpose—and know that you are not a random element in this world, because nothing that exists is random. You are an intrinsic part of this larger world, and as such, your life has purpose. And by "purpose," I mean the kind of purpose anchored in love and contributing to the health of the whole.

I extend this invitation as well to all the people who have lived in ways that have caused human suffering, as some of the characters in my father's stories lived. When we complain about our current state of affairs, we often lament "the system." As you may remember, however, each of my father's stories turned on the decisions of *individuals*. I see this as hopeful. Each one of us is born a free agent, and at any point in time we can choose to go deeper, to understand who we truly are, and then live accordingly.

I sometimes get distressed when I look out at a world that seems to be spinning out of control. I don't believe, however, that the purpose of our universe is to birth life that eventually extinguishes itself. I believe it's the existential challenge we now face that will motivate us to take a huge leap forward in our consciousness. Perhaps this is a test that is an intrinsic, required step in the evolution of consciousness throughout our universe—a kind of law of Nature—that forces life to shift from a linear, reductionist, and often self-serving way of thinking, to a heart-centered, healing, wholistic perspective. As the free-agent takers of this test, we can't assume we'll inevitably get there, but I'm an optimist.

As I said in the ending of my first film, *PlantPure Nation*, "The future is bright; all we need to do is see it." There is darkness in our world, but there is also tremendous opportunity for change. I truly would not want to be alive at another point in human history.

I hope you feel the same way.

If you are interested in helping to build a healthier, more connected, and kinder world, please visit NelsonCampbell.com.

There, you can stay connected with Nelson and Kim's ongoing work and learn about opportunities to engage in bottom-up efforts aimed at healing our world, one community at a time.

NELSONCAMPBELL.COM

ACKNOWLEDGMENTS

I must acknowledge my family, several of whom developed projects of various kinds that contributed significantly to this journey.

Unquestionably, this book—and in fact, all my books—would not have happened without the contributions of my wife of sixty-two years, Karen. In addition to constantly supporting me in every way imaginable, she also insisted I first write a book for the public, *The China Study*. Since then, she's been a constant presence in the books I have authored.

My children have also worked extensively to share my discoveries about nutrition with the public. My son Tom (BS, Cornell; MD, Univ. Buffalo), now a practicing physician who works with his wife, Erin (BS, MD, SUNY at Buffalo), a board certified preventive medicine physician, co-wrote *The China Study* with me, and later his own book, *The Campbell Plan*. My son Nelson (BA, MA, Cornell) produced two feature documentary films, *PlantPure Nation*, which documents how a state legislature (the legislature of Kentucky) officially responded to my nutritional research, and *From Food to Freedom*, which has received ten awards and honors at film festivals around the world. He also launched an international network of plant-based support and advocacy groups, a nonprofit organization, and a food and education business. Nelson's wife,

Kim (BS, Cornell), is an accomplished chef, with three acclaimed cookbooks (*The PlantPure Nation Cookbook*, *The PlantPure Kitchen*, and *PlantPure Comfort Food*), and has developed plant-based foods for mass production while also delivering educational programming. My daughter LeAnne (BS, MS, Cornell; PhD, UNC) was a Peace Corps volunteer in the Dominican Republic and published *The China Study Cookbook*. She currently serves as the CEO of the Center for Nutrition Studies.

I'd also like to thank those in our immediate family who, though not directly involved with these projects, are an important part of our support network. My son Keith is an accomplished, intelligent software engineer who has also become a skilled gardener, and his patient and loving wife, Eva, works as an IT security consultant. My other son, Dan, has a heart of gold and works as a special education teacher in a nearby public school, while his calm and centered wife, Lisa, works for a payroll processing company.

And finally, Karen and I are fortunate to have eleven grandchildren and two great-grandchildren, who have given us endless hours of joy and remind us why it's so important to do our best to build the future our young people deserve.

—Colin

I would like to begin by acknowledging my family, starting with the person I met when I was just sixteen—my wife, Kim. She has been my life partner in every sense: raising our three children with me and walking beside me through every other chapter of our lives, including the journey I shared in this book.

I am also grateful for our three children—Whitney, Colin,

and Laura—and for our son-in-law, Will, who married Laura. Together they are raising our first grandchild, a beautiful little boy who reminds me every day why we must all work to leave the world better than we found it.

And of course, I also want to acknowledge my parents. I will always cherish the opportunity I had to work with them on this book—I love them, along with the rest of my family, more than words can express.

Finally, I want to thank the many people who have, in one way or another, joined Kim and me on the challenging and unpredictable journey we began more than fifteen years ago. This includes the passionate employees of our organizations, volunteers, investors and other donors, consultants, outside partners, and so many others. At times the challenges and the stress have felt almost unbearable, and without the people I often think of as the "angels" in my life, I'm not sure we could have continued.

Despite the hardships, we have been deeply blessed—with an enduring marriage, a growing family, and the many wonderful people we have met and worked with along the way.

—Nelson

NOTES

1 “New International Study: U.S. Health System Fails Many Americans; Ranks Lowest on Health Equity, Access, and Outcomes,” The Commonwealth Fund, September 19, 2024, https://www.commonwealthfund.org/press-release/2024/new-international-study-us-health-system-fails-many-americans-ranks-lowest.

2 T. C. Campbell and L. Friedman, “Chemical Assay and Isolation of Chick Edema Factor in Biological Materials,” *Journal of Association of Official Analytical Chemists* 49, no. 4 (August 1, 1966): 824–828, https://doi.org/10.1093/jaoac/49.4.824.

3 K. Sargeant, Ann Sheridan, J. O’Kelly, et al., “Toxicity Associated with Certain Samples of Groundnuts,” *Nature* 192, no. 4807 (December 1961): 1095–1096, https://doi.org/10.1038/1921096a0.

4 Gerald. N. Wogan, “Aflatoxin Carcinogenesis,” in *Methods in Cancer Research, Volume 7*, ed. Harris Busch (Academic Press, 1973), 309–344.

5 T. C. Campbell, “Report to the USAID Mothercraft Center Project” (report, Virginia Polytechnic Institute and State University and Philippine Department of Health, 1967).

6 T. V. Madhavan and C. Gopalan, “The Effect of Dietary Protein on Carcinogenesis of Aflatoxin,” *Archives of Pathology & Laboratory Medicine* 85, no. 2 (February 1968): 133–137.

7 G. J. Mulder, *The Chemistry of Vegetable & Animal Physiology*, trans. P. F. H. Fromberg (W. Blackwood & Sons, 1849); Hubert Bradford Vickery, “The Origin of the Word Protein,” *Yale Journal of Biology and Medicine* 22, no. 5 (May 1950): 387–393, https://pmc.ncbi.nlm.nih.gov/articles/PMC2598953/.

8 T. C. Campbell, J. K. Loosli, R. G. Warner, et al., “Utilization of Biuret by Ruminants,” *Journal of Animal Science* 22, no. 1 (February 1, 1963): 139–145, https://doi.org/10.2527/jas1963.221139x; T. C. Campbell, R. G. Warner, J. K. Loosli, in “Cornell Nutrition Conference” (conference proceedings, Buffalo, NY, 1960), 96–103.

9 T. Colin Campbell Center for Nutrition Studies, https://nutritionstudies.org/.

10 T. V. Madhavan and C. Gopalan, “The Effect of Dietary Protein on Carcinogenesis of Aflatoxin,” *Archives of Pathology & Laboratory Medicine* 85, no. 2 (February 1968): 133–137.

11 J. Chen, M. P. Goetchius, G. F. Combs Jr., et al., "Effects of Dietary Selenium and Vitamin E on Covalent Binding of Aflatoxin to Chick Liver Cell Macromolecules," *Journal of Nutrition* 112, no. 2 (February 1982): 350–355, doi: 10.1093/jn/112.2.350; J. Chen, M. P. Goetchius, T. C. Campbell, et al., "Effects of Dietary Selenium and Vitamin E on Hepatic Mixed-Function Oxidase Activities and In Vivo Covalent Binding of Aflatoxin B1 in Rats," *Journal of Nutrition* 112, no. 2 (February 1982): 324–331, doi: 10.1093/jn/112.2.324.

12 J. Y. Li, B. Q. Liu, G. Y. Li, et al., "Atlas of Cancer Mortality in the People's Republic of China. An Aid for Cancer Control and Research," *International Journal of Epidemiology* 10, no. 2 (June 1981): 127–133, doi: 10.1093/ije/10.2.127.

13 J. Chen, T. C. Campbell, J. Li, et al., *Diet, Life-Style and Mortality in China: A Study of the Characteristics of 65 Chinese Counties* (Oxford University Press, Cornell University Press, People's Medical Publishing House, 1990).

14 Jane E. Brody, "Huge Study of Diet Indicts Fat and Meat," *New York Times*, May 8, 1990, https://www.nytimes.com/1990/05/08/science/huge-study-of-diet-indicts-fat-and-meat.html.

15 T. Colin Campbell, with Nelson Disla, *The Future of Nutrition* (BenBella Books, 2020).

16 D. Armstrong and R. Doll, "Environmental Factors and Cancer Incidence and Mortality in Different Countries, with Special Reference to Dietary Practices," *International Journal of Cancer* 15 (1975): 617–631.

17 D. Ganmaa and A. Sato, "The Possible Role of Female Sex Hormones in Milk from Pregnant Cows in the Development of Breast, Ovarian and Corpus Uteri Cancers," *Medical Hypotheses* 65: 1028–1037,

18 K. K. Carroll, L. M. Braden, J. A. Bell, and R. Kalamegham, "Fat and Cancer," *Cancer* 58 (1981): 1818–1825.

19 D. Armstrong and R. Doll, "Environmental Factors and Cancer Incidence and Mortality in Different Countries, with Special Reference to Dietary Practices," *International Journal of Cancer* 15 (1975): 617–631.

20 W. E. Connor and S. L. Connor, "The Key Role of Nutritional Factors in the Prevention of Coronary Heart Disease," *Preventative Medicine* 1 (1972): 49–83.

21 D. Armstrong and R. Doll, "Environmental Factors and Cancer Incidence and Mortality in Different Countries, with Special Reference to Dietary Practices," *International Journal of Cancer* 15 (1975): 617–631.

22 B. J. Abelow, T. R. Holford, and K. L. Insogna, "Cross-Cultural Association Between Dietary Animal Protein and Hip Fracture: A Hypothesis," *Calcified Tissue International* 50, no. 1 (1992): 14–18.

23 C. Cooper, G. Campion, and L. J. Melton, "Hip Fractures in the Elderly: A Worldwide Projection," *Osteoporosis International* 2, no. 6 (1992): 285–289.

24 BBC, "Statin-Fortified Drinking Water in the UK?" *Wastewater Digest*, August 4, 2004. Accessed November 15, 2025.

25 Paula Byrne, Maryanne Demasi, Mark Jones, et al., "Evaluating the Association Between Low-Density Lipoprotein Cholesterol Reduction and Relative and Absolute Effects of Statin Treatment: A Systematic Review and Meta-analysis," *JAMA Internal Medicine* 185, no. 5 (March 14, 2022): 474–481, doi: 10.1001/jamainternmed.2022.0134.

26 "Global Statin Market Size, Share, and Trends Analysis Report – Industry Overview and Forecast to 2032," Data Bridge Market Research, April 2025, https://www.databridgemarketresearch.com/reports/global-statin-market.

27 W. E. Connor and S. L. Connor, "The Key Role of Nutritional Factors in the Prevention of Coronary Heart Disease," *Preventative Medicine* 1, no. 1 (March 1972): 49–83, doi: 10.1016/0091-7435(72)90077-1; N. Jolliffe and M. Archer, "Statistical Associations Between International Coronary Heart Disease Death Rates and Certain Environmental Factors," *Journal of Chronic Diseases* 9, no. 6 (June 1959): 636–652, doi: 10.1016/0021-9681(59)90114-6.

28 "Type 2 Diabetes Market Size, Share & Segmentation by Drug Class [Insulin, DPP-4 Inhibitors, GLP-1 Receptor Agonists, SGLT2 Inhibitors, Others], by Route of Administration [Oral, Subcutaneous, Intravenous], by Distribution Channel [Retail Pharmacies, Hospital Pharmacies, Other], by Region and Global Forecast for 2024–2032," S&S Insider, March 2025, https://www.snsinsider.com/reports/type-2-diabetes-market-5955.

29 R. B. Shekelle, M. Lepper, S. Liu, et al., "Dietary Vitamin A and Risk of Cancer in the Western Electric Study," *Lancet* 2, no. 8257 (November 28, 1981): 1185–1190, doi: 10.1016/s0140-6736(81)91435-5; G. S. Omenn, G. E. Goodman, M. D. Thornquist, et al., "Effects of a Combination of Beta Carotene and Vitamin A on Lung Cancer and Cardiovascular Disease," *New England Journal of Medicine* 334, no. 18 (May 2, 1996): 1150–1155, doi: 10.1056/NEJM199605023341802.

30 Research America, "U.S. Investments in Medical and Health Research and Development, 2013–2018," https://www.researchamerica.org/wp-content/uploads/2022/09/InvestmentReport2019_Fnl.pdf.

31 T. Colin Campbell, with Howard Jacobson, *Whole: Rethinking the Science of Nutrition* (BenBella Books, 2013); T. Colin Campbell, with Nelson Disla, *The Future of Nutrition* (BenBella Books, 2020).

32 More information about this affair can be found on pp. 252–258 of *The China Study: Revised and Expanded Edition* by T. Colin Campbell and Thomas M. Campbell II (BenBella Books, 2016).

33 B. S. Appleton and T. C. Campbell, "Dietary Protein Intervention During the Post-Dosing Phase of Aflatoxin B1-Induced Hepatic Preneoplastic Lesion Development," *Journal of the National Cancer Institute* 70, no. 3 (March 1983) 547–549, https://pubmed.ncbi.nlm.nih.gov/6132019/; B. S. Appleton and T. C. Campbell, "Inhibition of Aflatoxin-Initiated Preneoplastic Liver Lesions by Low

Dietary Protein," *Nutrition and Cancer* 3, no. 4 (1982): 200–206, doi: 10.1080/01635588109513723.

34 Morgan Stanley Research, "Scaling Up the Impact of Obesity Drugs," Morgan Stanley, May 7, 2024.

35 T. Colin Campbell, "British Broadcasting Corporation (BBC), Your Credibility Is Tarnished," T. Colin Campbell Center for Nutrition Studies, January 25, 2017, https://nutritionstudies.org/british-broadcasting-corporation-bbc-your-credibility-is-tarnished/; Caldwell Esselstyn, Jr., "British Broadcasting Corporation (BBC), Your Credibility Is Tarnished Part 2," T. Colin Campbell Center for Nutrition Studies, January 27, 2017, https://nutritionstudies.org/british-broadcasting-corporation-bbc-credibility-tarnished-part-2/.

36 "Networks," Max Delbrück Center for Molecular Medicine in the Helmholtz Association, accessed October 28, 2025, https://www.mdc-berlin.de/research/themes/translation/ecrc/research_teams/networks.

37 PLANT BASED NEWS, "FAT GENES vs PLANT BASED DIET w/ Dr. Giles Yeo & Dr. T. Colin Campbell," YouTube, December 14, 2017, video, 21:42, https://www.youtube.com/watch?v=wnS8VsuchYY.

38 Committee on Diet Nutrition and Cancer, *Diet, Nutrition and Cancer.* (National Academy Press, Washington, DC, 1982), pp. 478.

39 Select Committee on Nutrition and Human Needs (U.S. Senate), "Dietary goals for the United States, 2nd Edition," (U.S. Government Printing Office, Washington, DC, 1977).

40 "Report Offers New Eating and Physical Activity Targets to Reduce Chronic Disease Risk," National Academy of Sciences, National Academies of Sciences, Engineering, and Medicine, September 5, 2002, https://www.nationalacademies.org/news/2002/09/report-offers-new-eating-and-physical-activity-targets-to-reduce-chronic-disease-risk.

41 "EatRightPRO.org," Academy of Nutrition and Dietetics, 2025, accessed October 28, 2025, www.eatrightpro.org.

42 "Conditions Contributing to Deaths Involving COVID-19, by Age Group, United States. Week Ending 2/1/2020 to 12/5/2020," National Center for Health Statistics, Centers for Disease Control and Prevention, December 9, 2020, https://www.cdc.gov/nchs/data/health_policy/covid19-comorbidity-expanded-12092020-508.pdf.

43 R. C. Bell, K. A. Golemboski, R. R. Dietert, et al., "Long-Term Intake of a Low-Casein Diet Is Associated with Higher Relative NK Cell Cytotoxic Activity in F344 Rats," *Nutrition and Cancer* 22, no. 2 (1994): 151–162, doi: 10.1080/01635589409514340.

44 T. Colin Campbell, "A Nutritional Link for COVID-19?," *EC Nutrition* 16, no. 2 (2021): 18–26, https://ecronicon.net/assets/ecnu/pdf/ECNU-16-00903.pdf.

45 Fiscal year 2020–2024 vaccine revenue and gross-profit disclosures (Pfizer 2024 Form 10-K; BioNTech 2024 Annual Report; Moderna 2024 Form 10-K; Johnson & Johnson 2024 Form 10-K; AstraZeneca 2024 Annual Report).

46 Annual COVID-19 vaccine revenue per firm, FY 2021 and FY 2022: Pfizer-BioNTech $36.8 bn (2021), $37.8 bn (2022); Moderna $17.7 bn (2021), $18.4 bn (2022); J&J $2.4 bn (2021), $2.8 bn (2022); AstraZeneca $4.97 bn (2021), $1.81 bn (2022). Totals ≈ $63 bn (2021) and ≈ $61 bn (2022). All figures from audited annual reports/10-K filings.

47 EvaluatePharma, *World Preview 2019, Outlook to 2024* (July 2019), corroborated by WHO, *Global Vaccine Market Report* (2020); "Pfizer's Vaccine Powerhouse Before Covid Was Prevnar. Here's Why It Still Matters," *Barron's*, Dec. 2, 2020.. Prevnar 13 sales peaked at around $6 bn in 2015 (Pfizer 2016 Form 10-K).

48 Visit https://foodstudies.org/ to learn more.

49 Beth Greenfield, "The Fascinating History of Baby Formula," *Yahoo Life,* August 23, 2022, https://www.yahoo.com/lifestyle/fascinating-history-of-baby-formula-231353714.html.

50 Jesse Anttila-Hughes, Lia Fernald, Paul Gertler, et al., "The Deadly Toll of Marketing Infant Formula in Low- and Middle-Income Countries," VoxDev, October 31, 2023, https://voxdev.org/topic/health/deadly-toll-marketing-infant-formula-low-and-middle-income-countries.

51 Michael Latham, press release, Cornell University, July 28, 1997.

52 Michael Latham, letter to Cutberto Garza, August 14, 1997, in the possession of T. Colin Campbell.

53 Andrew Jacobs, "Opposition to Breast-Feeding Resolution by U.S. Stuns World Health Officials," *New York Times,* July 8, 2018, https://www.nytimes.com/2018/07/08/health/world-health-breastfeeding-ecuador-trump.html.

54 Darren Peter Morton, Paul Rankin, Lillian Kent, et al., "The Complete Health Improvement Program (CHIP) and Reduction of Chronic Disease Risk Factors in Canada," *Canadian Journal of Dietetic Practice and Research* 75, no. 1 (July 2014): 72–77, doi: 10.3148/75.2.2014.72.

55 Jeffrey M. Jones, "In U.S., 4% Identify as Vegetarian, 1% as Vegan," Gallup, August 24, 2023, https://news.gallup.com/poll/510038/identify-vegetarian-vegan.aspx.

56 "How Many Former Vegetarians and Vegans Are There?," Faunalytics, December 2, 2014, https://faunalytics.org/how-many-former-vegetarians-and-vegans-are-there/.

57 P. J. Gerber, H. Steinfeld, B. Henderson, et al., *Tackling Climate Change Through Livestock – A Global Assessment of Emissions and Mitigation Opportunities*, Food and Agriculture Organization of the United Nations, 2013, https://www.fao.org/4/i3437e/i3437e.pdf.

58 Xu, Xiaoming, Prateek Sharma, Shijie Shu, Tzu-Shun Lin, Philippe Ciais, Francesco N. Tubiello, Pete Smith, Nelson Campbell, and Atul K. Jain, "Global

Greenhouse Gas Emissions from Animal-Based Foods Are Twice Those of Plant-Based Foods," Nature Food 2, no. 9 (September 2021): 724–32. https://doi.org/10.1038/s43016-021-00358-x.

59 Jean-François Bastin, Yelena Finegold, Claude Garcia, et al., "The Global Tree Restoration Potential," *Science* 365, no. 6448 (July 5, 2019): 76–79, doi: 10.1126/science.aax0848.

ABOUT THE AUTHORS

T. Colin Campbell, PhD, has been dedicated to the science of human health for more than sixty years. His primary focus is on the association between diet and disease, particularly cancer. Known for the China study—one of the most comprehensive studies of health and nutrition ever conducted, and recognized by *The New York Times* as the "Grand Prix of epidemiology"—Dr. Campbell's profound impact also extends to education, public policy, and laboratory research. Dr. Campbell is the Jacob Gould Schurman Professor Emeritus of Nutritional Biochemistry at Cornell University. He has received more than seventy grant years of peer-reviewed research funding and authored more than three hundred fifty research papers. The China study was the culmination of a twenty-year partnership of Cornell University, Oxford University, and the Chinese Academy of Preventive Medicine.

On the heels of writing and directing the 2015 documentary film *PlantPure Nation*, **Nelson Campbell** founded the nonprofit PlantPure Communities (PPC) in early 2016. Released in theaters in over one hundred cities, the *PlantPure Nation* film (now on Amazon Prime and Apple TV) highlights the dramatic healing power of a plant-based diet and examines the political and economic factors that have suppressed information about these healing benefits, while making connections to public policy, medical practice, food deserts, and farming.

After the release of *PlantPure Nation*, PPC launched and supported an international network of local support and advocacy groups called "Pods," engaging hundreds of thousands of people. This network was handed off to the Physicians Committee for Responsible Medicine in 2023 to enable PPC to focus on media-driven strategies for sharing its health message and on outreach in under-resourced communities.

PPC's most recent media production is a second documentary titled *From Food to Freedom*, which Nelson also wrote and directed. This film tells a dramatic story demonstrating the power of nutrition to rapidly heal people with type 2 diabetes. It won ten awards at film festivals around the world, including awards for Best Feature Documentary (SoCal Film Awards) and Best Inspiring Film (LA Film Awards).

In addition to his work in the nonprofit sector, Nelson founded the PlantPure foods and education business. PlantPure has been working on a strategy in western North Carolina to partner with local communities to share the healing benefits of plant-based nutrition. This strategy also includes a commitment to waive 100 percent of its profit margin on sales of certain of its products in under-resourced communities.

Most recently, Nelson has launched a Substack publication at NelsonCampbell.com, where he writes about many of the same issues shared in *The Whole Truth*.

Finally, Nelson has also had extensive experience in public speaking, mostly focused on issues surrounding plant-based nutrition, but also touching on public policy, economics, and social concerns.

Prior to his involvement with *PlantPure Nation*, Nelson worked for twenty-five years as a business entrepreneur. He graduated from Cornell University with a bachelor's degree in Political Science and a master's degree in Economics.